Zimbabwe's Migrants and So
Border Farms

The Roots of Impermanence

During the Zimbabwean crisis, millions crossed through the apartheid-era border fence, searching for ways to make ends meet. Maxim Bolt explores the lives of Zimbabwean migrant labourers, of settled black farm workers and their dependents, and of white farmers and managers, as they intersect on the border between Zimbabwe and South Africa. Focusing on one farm, the book investigates the role of a hub of wage labour in a place of crisis. A close ethnographic study, this book addresses the complex, shifting labour and life conditions in northern South Africa's agricultural borderlands. Underlying these challenges are the Zimbabwean political and economic crisis of the 2000s and the intensified pressures on commercial agriculture in South Africa following liberalisation and post-apartheid land reform. But, amidst uncertainty, farmers and farm workers strive for stability. The farms on South Africa's margins are centres of gravity, islands of residential labour in a sea of informal arrangements.

Maxim Bolt is Lecturer in Anthropology and African Studies at the University of Birmingham and a Research Associate at the Wits Institute for Social and Economic Research, University of the Witwatersrand. His doctoral thesis, completed after the research on which this monograph draws, was awarded runner-up in the biennial Audrey Richards Prize by the African Studies Association of the UK. He has published in the leading African Studies and anthropology journals, and serves on the editorial boards of *Africa* and the *Journal of Southern African Studies*.

THE INTERNATIONAL AFRICAN LIBRARY

General Editors

J. D. Y. PEEL, *School of Oriental and African Studies, University of London*
LESLIE BANK, *Fort Hare Institute of Social and Economic Research, South Africa*
HARRI ENGLUND, *University of Cambridge*
DEBORAH JAMES, *London School of Economics and Political Science*
ADELINE MASQUELIER, *Tulane University*

The International African Library is a major monograph series from the International African Institute. Theoretically informed ethnographies, and studies of social relations 'on the ground' which are sensitive to local cultural forms, have long been central to the Institute's publications programme. The IAL maintains this strength and extends it into new areas of contemporary concern, both practical and intellectual. It includes works focused on the linkages between local, national, and global levels of society; writings on political economy and power; studies at the interface of the socio-cultural and the environmental; analyses of the roles of religion, cosmology, and ritual in social organisation; and historical studies, especially those of a social, cultural, or interdisciplinary character.

For a list of titles published in the series, please see the end of the book.

Zimbabwe's Migrants and South Africa's Border Farms

The Roots of Impermanence

Maxim Bolt
University of Birmingham

International African Institute, London
and
CAMBRIDGE UNIVERSITY PRESS

CAMBRIDGE
UNIVERSITY PRESS

University Printing House, Cambridge CB2 8BS, United Kingdom

One Liberty Plaza, 20th Floor, New York, NY 10006, USA

477 Williamstown Road, Port Melbourne, VIC 3207, Australia

4843/24, 2nd Floor, Ansari Road, Daryaganj, Delhi - 110002, India

79 Anson Road, #06-04/06, Singapore 079906

Cambridge University Press is part of the University of Cambridge.

It furthers the University's mission by disseminating knowledge in the pursuit of education, learning and research at the highest international levels of excellence.

www.cambridge.org
Information on this title: www.cambridge.org/9781107527836

© Maxim Bolt 2015

This publication is in copyright. Subject to statutory exception and to the provisions of relevant collective licensing agreements, no reproduction of any part may take place without the written permission of Cambridge University Press.

First published 2015
First paperback edition 2017

A catalogue record for this publication is available from the British Library

Library of Congress Cataloging in Publication data
Bolt, Maxim, 1970– author.
Zimbabwe's migrants and South Africa's border farms : the roots of impermanence / Maxim Bolt, University of Birmingham.
  pages cm. – (The international African library)
Includes bibliographical references and index.
1. Migrant agricultural laborers – South Africa – Social conditions. 2. Foreign workers, Zimbabwean – South Africa. 3. Farmers – South Africa. 4. Borderlands – South Africa. 5. Borderlands – Zimbabwe. 6. South Africa – Race relations – 21st century. I. Title. II. Series: International African library.
HD5856.S6B65 2015
331.5440968 – dc23   2015015331

ISBN 978-1-107-11122-6 Hardback
ISBN 978-1-107-52783-6 Paperback

Cambridge University Press has no responsibility for the persistence or accuracy of URLs for external or third-party internet websites referred to in this publication, and does not guarantee that any content on such websites is, or will remain, accurate or appropriate.

For Oma

Contents

Maps and Figures	*page*	xi
Acknowledgements		xiii
List of Key Characters		xvii
1 Introduction: Labour and Fragmentation on the Limpopo River		1
2 'It's in Our Blood, It's in Our Skin': Success, Failure, and Self-Sufficiency in Border Farming		38
3 Behind the Mountain: Core, Periphery, and Control in the Limpopo Valley		67
4 Producing Permanence: Employment and Domesticity in the Black Workforce		105
5 Reimagining Men: Middle-Class Farm Workers and the Zimbabwean Crisis		137
6 'Management' or 'Paternalism'?: Race and Registers of Labour Hierarchy		159
7 Scaling Up: The Farms and the Border Economy		182
8 Conclusion: Between Production and Fragmentation		207
References		219
Index		235

Maps and Figures

Maps

1	The Zimbabwean–South African border in the region	*page* xix
2	The Soutpansberg and Limpopo Valley	xix

Figures

2.1	The border farming area	48
3.1	Messina mining camp in 1907	80
3.2	Early shaft sinking at Messina	81
3.3	The new mill, completed in 1913	82
3.4	Planned expansion of Messina Health Committee's jurisdiction	84
4.1	*Mapermanent* workers supervising the picking operation	111
4.2	Grootplaas packshed, showing gantries and personnel manager's office	113
4.3	The Grootplaas labour compound (the New Houses)	117
4.4	The New Houses, quiet during working hours	117
4.5	Seasonal workers sit around a fire in the New Houses	119
4.6	Inside a New Houses room	120
4.7	The Hostel	121
4.8	*Mapermanent* in a yard constructed outside a room	126
5.1	When the harvest begins: job seekers are kept out of the workshop yard by security guards and supervisors	138
5.2	Picking team	146
5.3	The object of avoidance: dancing and drinking at the New Houses	155
6.1	Marula and André gather pickers for a meeting	166
6.2	The yard of Marula's house, with the *khoro* on the right	170
6.3	Michael's yard	172
7.1	Park Station, the compound car park	194
7.2	Informal *bakkie* taxi carrying departing seasonal workers and their supplies	195
7.3	Monthly payday market in Grootplaas compound	200

7.4 Seasonal workers leaving the compound after the
    harvest, at dawn   201
7.5 Seasonal workers waiting with luggage at the border,
    after a trip to Musina   202

Acknowledgements

This book emerged from a PhD thesis, which I undertook at the London School of Economics. For help developing both the thesis and this book, I owe a lot of people a great deal of gratitude. The farming family of the estate I call Grootplaas allowed me to reside on their land for 17 months, and I owe them enormous thanks for their hospitality. I owe equal thanks to the farm workers and residents with whom I lived, for taking me in, looking after me untiringly, and assisting me with my research. In both cases, protecting their identities prevents me from mentioning them by name. For helping prepare for fieldwork by putting me in touch with farmers' associations, thanks to Blair Rutherford, and thank you in turn to the officials of AgriLimpopo farmers' association, Amanda James and Gert Rall, who helped me contact farmers. Thank you also to Jack Klaff and Brian Kalshoven for allowing me to use their photographs and archived material on Musina's local history, and to Colbert Tshivhase for introducing me to different places around the former homeland of Venda, and for translating recorded interviews. I am deeply grateful to Fiona Nicholson, Fliss Ingham, and Sonia Rupcic for being there for me throughout my fieldwork, and to Fraser McNeill for the years of guidance and support, both in South Africa and in London. In London, Chenjerai Shire has been a wonderful teacher of ChiShona and a provocative interlocutor.

    I would probably not have ended up where I did had it not been for Gavin Williams, my undergraduate tutor at St Peter's College, Oxford. Although I did not know it yet when he taught me, his interests and perspectives have doubtless made their mark. At the London School of Economics, where I was generously supported by an Economic and Social Research Council Studentship (Award PTA-031-2005-00006), I benefited from being surrounded by an amazingly nurturing group of colleagues. They are too numerous to mention, but I would especially like to thank Laura Bear, Judith Bovensiepen, Tom Boylston, Vicky Boydell, Fenella Cannell, Alanna Cant, Sharad Chari, Ankur Datta, Amit Desai, Elizabeth Frantz, Chris Fuller, Ana Paola Gutierrez Garça, Carrie Heitmeyer, Michael Hoffmann, Elizabeth Hull, Malcolm James, Nicolas Martin, Aude Michelet, Martha Mundy, Michelle Obeid, Dave

Robinson, Andrew Sanchez, Charles Stafford, Hans Steinmüller, George St Clair, Jason Sumich, Anna Tuckett, and Matt Wilde for helping me work through ideas and arguments, reading and commenting on my work, and even listening to trial runs of seminar papers. More generally, thank you to all members of the writing-up seminars in which I participated, for what was always insightful and positive engagement. Thanks to Yanina Hinrichsen, who took countless logistical problems on her own shoulders, and ensured that everything during my PhD was (relatively) smooth and painless. Charlotte Bruckermann, Thomas Grisaffi, Jason Hickel, Insa Koch, and Alpa Shah only joined the LSE after I had left, but long, wide-ranging discussions with all of them have been invaluable. Beyond the LSE, working with Jan-Bart Gewald and Paul Nugent at the AEGIS Summer School, as soon as I returned from fieldwork, was enormously helpful in setting me on the right path. Meanwhile, Amanda Hammar, Loren Landau, and JoAnn McGregor helped me work through ideas at an early stage of writing up, in the form of an article. Though always at different institutions, Zoe Groves and I have been in step with each other since our PhDs, and I hope our projects continue to intersect and produce great conversations and collaborations.

I developed the ideas in the thesis further during a postdoctoral fellowship at the British Museum, and most recently as a lecturer in the Department of African Studies and Anthropology at the University of Birmingham and a research associate at the Wits Institute of Social and Economic Research, University of the Witwatersrand. All of these have been exciting intellectual environments. I really appreciate the months thrashing out the relative merits of economic anthropology and economic history with Leigh Gardner in our office alcove in the British Museum. At DASA in Birmingham, I have been very lucky to have such encouraging colleagues – Karin Barber, Stewart Brown, Reg Cline-Cole, Insa Nolte, Katrien Pype, Benedetta Rossi, Keith Shear, and Kate Skinner – while Sadiah Qureshi upstairs in History has taken me under her wing and looked out for my best interests. Spending time at WISER has given me the opportunity to push ideas in fresh directions – especially with Keith Breckenridge, as far as the arguments in this book are concerned.

I am, of course, particularly indebted to Deborah James, unfailingly understanding and meticulous, and Matthew Engelke, always an inspirational mentor. As PhD supervisors and later as colleagues, both have offered me a vast amount of support and constructive advice over the years. I am similarly grateful to my PhD examiners, Jane Guyer and Jonathan Parry, for their stimulating and encouraging guidance – again, extending beyond the PhD – and to Amanda Hammar and Blair Rutherford, for their careful reading of the book manuscript and their thought-provoking comments and suggestions. Thank you also to Stephanie Kitchen at the International African Institute, for taking me through the whole process of preparing a book manuscript for publication, to Mike Kirkwood for copyediting, to Miles Irving for preparing the maps and illustrations, and to Ed Emery for preparing the index.

Finally, all of this has been made possible by the constant, loving encouragement of my parents, Neville and Lisa; my sister Olivia and her partner Matt; my grandmother Alexa; and a number of close friends outside the world of anthropology. Through this whole saga, my partner Jess has kept things in perspective, reading (almost) everything I have written, and even returning to my fieldsite with me after my PhD to catch up with friends and informants.

Needless to say, any errors or shortcomings are my own.

Earlier versions of Chapters 5 and 7 appeared as articles in the *Journal of Southern African Studies* (Bolt 2010) and *Africa* (Bolt 2012). An article in *Economy and Society* (Bolt 2013) drew on data presented in Chapters 2 and 4.

List of Key Characters

The Grootplaas Farming Family

Koos:	founder of Grootplaas, now retired
Willem:	Koos's son-in-law, who now runs Grootplaas
Jacques:	Willem's son, who manages an estate up the road
Paul:	Koos's son, who until recently helped Willem run Grootplaas

The White Border Farmers and Managers

Jan:	Koos's neighbour, and farming partner in the 1980s
Thinus:	one of three brothers on the border, and producer of tomatoes for ketchup
Jim:	Thinus's partner in one of his land portions
Dirk:	formerly a border farmer, who went bankrupt. Now Jan's manager
André:	the Grootplaas production manager

The Grootplaas Workforce and Compound Residents

Michael:	personnel manager
Holly:	Michael's partner
Marula:	foreman
Benjamin:	storeman
Hardship, Ezekiel:	members of Marula's Lands Team and harvest-time supervisors
Norman:	senior driver
Sarah:	Norman's wife
Joyce:	Norman's partner at the farm
Margaret:	former teacher, and the wife of a security guard
Alex, Vusa, Simon, Tendai:	young, educated Zimbabwean men working as pickers
George:	educated Rastafarian, whose shop was destroyed in *Murambatsvina*, now a picker

Chipo:	cigarette salesman and picker
Jameson:	former teacher, with university education, now a picker
Jenny:	Jameson's wife
Daniel:	tractor driver
MaiJimmy:	Daniel's sister, a semi-permanent worker who runs a *shebeen*
Josiah:	owner of a rival *shebeen*
Cornelius:	taxi driver

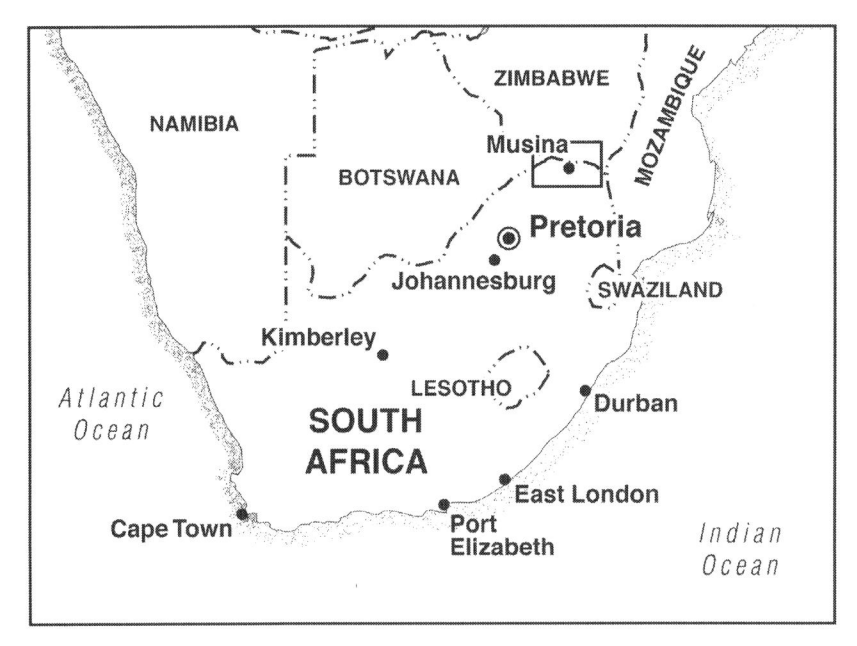

Map 1. The Zimbabwean–South African border in the region.

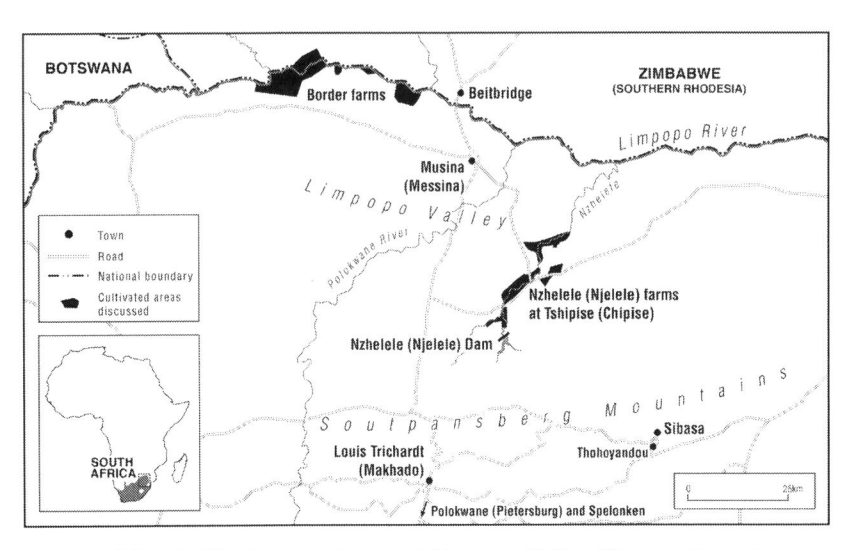

Map 2. The Soutpansberg and Limpopo Valley. (Previously used names are given in brackets.)

1 Introduction
Labour and Fragmentation on the Limpopo River

South Africa's great spinal road, the N1, runs from Cape Town north to Zimbabwe. Past Johannesburg, Pretoria, and hundreds of miles of green Highveld; over the subtropical ridge of the Soutpansberg at Louis Trichardt; through the flat, dry, scrubby *mopaneveld*[1] of the Limpopo Valley, and the bustle, trade, and exhaust fumes of Musina town. Then it collides abruptly with the border, near South Africa's northernmost point. Following the Limpopo River westwards, rather than crossing into Zimbabwe, means following a winding, potholed military service road. The road, in turn, tracks an endless serpent of razor wire coils, the centrepiece in an undulating strip of sandy no man's land. This defensive line, punctuated by garrisons, was erected during the 'border war' of apartheid's twilight. Its electric wires, once lethal, are now set to 'detect'. The patchwork of repairs, and the ceramic parts strewn across the ground, show that the fence is not as formidable as it might seem.

Across the road from the fence, dense riverside vegetation is punctuated by human settlement. Gaps in the foliage expose the tomato fields, cotton plantations, and citrus orchards of white-owned farms, and the regimented brick blocks or ramshackle mud houses of their labour compounds. The occasional *bakkie* (pick-up truck) – an informal taxi driver, a white farmer, a black foreman, a police or military patrol – shuttles between estates, garrisons, and the border post and town to the east. But the spaces between sites of settlement are quiet. A turn in the road, and there is not a soul to be seen.

The emptiness is vital for people hoping to cross the dry Limpopo riverbed undetected, climb and cut through the fence, and make their way southwards through the bush. A sparsely populated landscape is also a danger. Game farmers find corpses slumped against trees in the silent, parched expanse of *mopaneveld* – their postures exhausted, their empty water bottles still in hand. The bush stretching back from the border is a sea surrounding islands of crop farms and their working populations.

Undocumented migrants generally try to stay out of sight – with good reason. So, at six o'clock one April morning in 2008, South African

---

[1] Bush dominated by *mopane* trees.

soldiers newly posted to the area were all the more perturbed by the sight of hundreds of 'border jumpers'. In the biting cold of dawn, they crowded outside a farm, just off the border road a stone's throw from the garrison. Conspicuously undeterred by the soldiers' presence, they were looking for work.

At Grootplaas Estates, recruitment for the harvest was under way. In the yard outside the farm's workshop, women who had been employed in previous years stood in rows. Some jostled; others rushed forward. All hoped to be selected first for jobs in the 'packshed', sorting and boxing the farm's oranges and grapefruits for export. The black foreman and his team of work supervisors struggled to maintain control. Even Willem, the usually aloof white farmer-landowner, in khaki two-tone shirt and shorts, came over to call for order.

The real fight would be at the yard's perimeter fence, where most job-seekers waited. As usual, cross-border networks had relayed the recruitment date at least as far as Beitbridge, the Zimbabwean border town some 50 kilometres to the east. The soldiers entered the yard, but the crowd of 'illegals' clearly saw the recruitment process as trumping any state presence. Willem approached them, his Alsatian at his heel. If the soldiers left his recruitment alone, they would have a list of workers' names by the end of the day – a gesture towards distinguishing between those undocumented Zimbabweans who were farm workers, and those who were simply 'border jumpers'.

For the moment, work seekers were neither border jumpers, nor farm workers. They were suspended between categories. For the harvest of 2008, and unlike previous years, the white farmers had in fact agreed that their senior black employees would compile recruitment lists beforehand. Thus, the struggle for employment had already largely been won or lost – though many did not know this – in earlier appeals to powerful members of the workforce. Many first-timers, as well as the usual faces, had made the cut. Even so, until the list was read, the soldiers had no idea who among the crowd would be employed, and who would be left as 'just' a transient, undocumented migrant.

As Marula,[2] the black foreman, and André, the white production manager, drove the short distance to the gates of the yard, the crowd surged forward, and was ordered back onto the dirt road outside. The soldiers decided to take control, pushing people back beyond the fence with their guns. Security guards made a show of joining in, but their efforts lacked any real conviction. With the gate now shut, Hardship – keeper of the

---

[2] Although evoking an alcoholic drink, this pseudonym reflects the character of the foreman's real name. His nickname-like designation appears officially in his national identity book, and belongs to a register of black South African names that speaks of a life residing and working on white farms.

employment register – began to call out names. One soldier peered over his shoulder, as though checking that the list was genuine. The crowd, stretching back down the dirt road to the border fence, fell silent as hopefuls strained to hear. Those chosen pushed their way forward and were let through the gate. Eventually, the recruitment roll was completed: two teams of 30 male pickers and one of 30 female pickers; and 40 men and 80 women for the packshed. There would be further recruitment, on top of the 210 just chosen and the 120 employed a few days before, to make up Grootplaas's usual complement of around 460 seasonal workers. But surprise and disappointment were unmistakeable among the huge crowd left outside the gate.

There was a more immediate concern. The line between workers and border jumpers had been drawn, for now. Those left in the crowd were simply 'illegals', incongruously arrayed in front of a South African military patrol. People scattered into the citrus orchards, chased briefly by a soldier before he tripped over his gun strap and dropped his magazine. Most would walk a few kilometres to the neighbouring estates, hoping for work there. Some would soon be back for the next round of recruitment at Grootplaas. Once the soldiers had gone, a few men and women approached supervisors and the white production manager. They complained that promises had not been fulfilled, trying one last time to find employment through quiet negotiation.

At the moment of recruitment, diverse people, with diverse hopes and plans, become divided into clear groups. On the one hand are those without employment. In the eyes of soldiers, these are simply 'border jumpers', transients fleeing Zimbabwe's crisis, with no claim on the border as a place of labour. On the other are 'workers', built around a core of permanent employees – 140 at Grootplaas – whose presence is legitimated by their attachment to the farm.

Many new arrivals do indeed live a fugitive existence. The Limpopo River presents risks: from drowning or crocodile attacks when it is in flood, to abuse, assault, or rape by *magumaguma* – gangs that operate along the border – or by South African soldiers (see SPT 2004; cases in Orner and Holmes 2011). Some arrivals on the farms lack basic resources for immediate survival. Money, mobile phones, and even South African contact numbers may be casualties to robbery (see Hall 2013). On the farms themselves, aggressive border policing leaves recruits vulnerable to deportation raids.

Migrants' experience of transience is one of temporal fragmentation and spatial insecurity. Contexts of crisis, characterised by 'a loss of coherence and unity' (Vigh 2008: 10), often mean little scope for planning for the future. In Zimbabwe (see Hammar 2014; Jones 2014), as elsewhere, fast-shifting, unpredictable circumstances require constant navigation, as personal projects collapse, and norms formerly taken for granted cease to produce intended effects (Vigh 2008; see also Mbembe and Roitman 1995). On the Zimbabwean-South African border, displacement and the search for employment foreground a distinctly spatial dimension. As migrants seek shelter and work, and as they attempt to avoid apprehension, connections to places are tenuous at best. Crisis and transience here mean not only living from day to day, but also looking over one's shoulder, and figuring out where to go next.

On the border farms, recruitment renders migrants' transience starkly visible. Yet, away from the ritual of recruitment, distinctions between people become more complex. The farms and their settled workforces represent sites around which people gravitate. Situated far from the closest South African town, the farm's apparent isolation and stability is belied by networked connectedness, mobility, and variety. The majority of workers hail from Zimbabwe, but many have relatives elsewhere in South Africa with whom they remain in constant contact. Large numbers of Zimbabwean seasonal workers are employed on the farms each winter, and while some stay for the whole harvest, others quickly move on southwards into South Africa. Others again, who lack employment, are connected to workers through kinship, friendship, and sexual relationships. The farms integrate and root people – albeit to different degrees, and in unequal and sometimes precarious ways. They do so by offering them incorporation into workforces, and futures in the relatively ordered worlds of labour compounds. Lives amidst crisis can themselves become routinised to a degree (Mbembe and Roitman 1995; Vigh 2008; Finnström 2008). But routine on the farms is much further-reaching. Here, wage labour organises time and work hierarchies organise people.

As migrants attempt to gain a foothold in something stable, the farms themselves are far from unchanging. This monograph begins with the story of the rise of the border farms in the 1980s, following the earlier ascendance of other forms of capitalist production in South Africa's far north. The farms were forged in the very crucible of mobility, as white farmers left newly independent Zimbabwe or came from other parts of South Africa to plant crops, and temporarily root themselves, on the southern bank of the Limpopo River. The wider geopolitics of the region, with its emphasis on redressive and restitutive measures, now in turn makes for an uncertain future. Farmers insure themselves against the possible vagaries of South African land reform by spreading risk across several farms, even securing land in Mozambique so that they can leave South Africa if necessary. Their expansion is equally a response to a liberalised market in which only the largest operators survive. At the same time, some have invited Australian coal prospectors onto their land, keeping their options open in case they are forced to sell. This is far from the popular view of sedentary farmers, rooted in land passed down the generations. For white farmers and black workers, everyday stability remains provisional.

And yet the estates have become focal points around which diverse residents organise their lives, amidst a bewildering kaleidoscope of economic informalisation and political turmoil. Being a farm worker, not a transient border jumper, means being incorporated into arrangements of stability, even permanence, however provisional. This book explores the lives of migrants, of settled black farm workers and their dependants, and of white farmers and managers, as they intersect on the border. Focusing on one farm, it investigates the role of a hub of wage labour in a place of crisis. *The Roots of Impermanence* argues that, for people facing uncertain futures in current regional upheavals and global capitalism, workplaces are lifeplaces.

Insecurity, Labour, and the Meaning of Mobility

This is one perspective on capitalism today. Contemporary flexible capitalism stands for ephemerality and perpetual change. Local arrangements are thought to be so *ad hoc* and fleeting that contracts collapse into informality, employment into entrepreneurialism. Acute crisis is seen merely to hasten capitalism along its path. The notoriously chaotic national border between unemployment-ridden South Africa and Zimbabwe, with its political and economic troubles, is certainly a place of crisis. But here local centres of capitalist production – the border's plantation estates – represent centres of gravity, islands of wage labour in a sea of informal arrangements.

The struggle for farm employment on the border is desperate and sometimes violent. This is all the more striking because agricultural labour is some of the least desirable work around. Understanding the significance of this book's argument – that workplaces and work relationships root people – means appreciating the acute insecurity that pertains in contemporary Zimbabwe and South Africa.

The scene above is an indicator of Zimbabwe's recent political and economic crisis. Since 2000, the crisis has precipitated one of the world's highest-ever rates of hyperinflation (see, for example, *Guardian* 2008) and severe supply shortages. This has led to the displacement of millions of people across the region in search of livelihoods. Most head for South Africa. Statistics are unreliable because most Zimbabweans come through the border fence but, by 2009, Doctors Without Borders put the number at an estimated total of 3 million (Doctors Without Borders 2009). In South Africa, a citizenry frustrated by mass unemployment has reacted with violent xenophobia (see Morreira 2010). In April 2008, when this book's opening scene took place, hyperinflation in Zimbabwe was escalating fast. In South Africa, deportations of Zimbabweans were rampant, and widespread attacks on people perceived to be foreigners were about to hit headlines. Against the backdrop of all this uncertainty, April also saw the first of Zimbabwe's controversial 2008 elections, an event to which I return later.

As with similar forms of instability elsewhere, such as the economic decline in the Democratic Republic of Congo that drove many 'respectable' people to live by fending for themselves (MacGaffey and Bazenguissa-Ganga 2000), a large number of Zimbabweans seek unfamiliar means, outside their home country, of making ends meet. Many people who pass through the border have only a vague sense of how they will do so. Those who do not find jobs on any of the farms keep going, southwards. Some have relatives in cities or in smaller towns. For others, Johannesburg beckons, with its famed opportunities. But the Central Methodist Church in the heart of the metropolis is a good indication of what awaits: at night, its winding corridors are packed, body-to-body, with Zimbabweans seeking shelter.[3]

These predicaments intersect with others in the region. The Zimbabwean crisis is only one among a cluster of processes of fragmentation. Opportunities for formal employment have contracted. South Africa and Zimbabwe, especially, previously drew vast numbers of black people from across Southern Africa into formal, though inequitable, employment as labour migrants. The 1990s saw both countries turning to leaner economic models. In today's South Africa, government policies have paved the way for neo-liberal open markets and 'flexible accumulation' (Harvey 1990). The widespread casualisation of work, in especially large numbers on commercial farms (Ewert and du Toit 2005; Addison 2006; Rutherford and Addison 2007), coexists with secure employment for the privileged few, who become targets of such initiatives as Corporate Social Responsibility (see, for example, Rajak 2008). Most people are excluded from such employment altogether. In 1990s Zimbabwe, similarly severe loss of employment followed 'the global imperatives of the structural adjustment programme' (Raftopoulos and Phimister 2004: 357; see also Gibbon 1995).

The 'informal economy', as practice and academic concern, now appears to sum up livelihood opportunities. Crisis in Zimbabwe meant 'informalisation' of employment, tenure rights, party politics, and migration (Raftopoulos, conference keynote quoted in Hammar, McGregor, and Landau 2010: 269). With the collapse of much Zimbabwean industry and commercial agriculture, ordinary people are reduced to hunting

---

[3] Bishop Paul Verryn's Central Methodist Church in downtown Johannesburg has, since 2004, been a *de facto* refuge for countless migrants, most from Zimbabwe. It consequently became a focus for high-profile humanitarian efforts. See Beremauro's recent ethnographic study (2013). In 2015 migrants have come under renewed pressure to leave the church and have been subject to deportation raids (see Guardian 2015; Al Jazeera 2015).

ad hoc, survivalist opportunities to make ends meet: *kukiya-kiya*, 'making do' by means of 'zigzag' arrangements (Jones 2010). For many in South Africa, finding an ever-smaller niche in the overcrowded world of small-scale trade appears the only option beyond state benefits (Ferguson 2007; see James and Hull 2012). In both countries, many people's livelihoods and projects have insecure, short-term horizons. The days of the long-term plan, achieved through stable employment, seem a remote memory: remitting throughout years of work, accumulating a fund for respectable retirement.

Changes in patterns of spatial mobility reflect – indeed, epitomise – this wider fragmentation. Large-scale migrant labour in centres of capitalist production appears to have disappeared overnight, together with the academic preoccupations that had accompanied it. Earlier scholarly focus on labour migrancy in southern Africa,[4] the 'productivist' historiography of 'the Africa of the labour reserve' (Andersson 2006: 376), has given way to trade networks (see Andersson 2006), new forms of debt and destitution (Hull and James 2012), and identity and 'abjection' in the neo-liberal world order (Ferguson 1999).

Yet hubs of waged, migrant labour have not simply disappeared. What does employment – now far scarcer – mean amidst such transience? What was at stake for the Zimbabweans at the Grootplaas gates? Answering these questions means thinking differently about labour migration itself. Labour migration and displacement – or 'forced migration' – are the subjects of two largely separate bodies of literature, with different guiding questions. Scholarship on the latter often explores the nature of people's attachment to places and their experiences of being uprooted.[5] Labour migration literature, on the other hand, assuming longer-term continuity, seeks to establish whether migrants seek material accumulation or personal transformation when they move,[6] and explores the moral struggles surrounding, and the consequences of, remittance patterns.[7]

In Southern Africa, too, labour migration and displacement have been considered as two different kinds of human mobility. Until apartheid ended in 1994, 'labour migration' was the central paradigm. Literature emphasised the integration of Zimbabwe, South Africa, and neighbouring nation states into a single regional political economy in which countless black people from across the subcontinent sought employment

---

[4] While this book takes Zimbabwe and South Africa as its particular focus, there have also been important hubs of labour migration in other countries such as Zambia (see, for example, Epstein 1958) and Namibia (see, for example, Gordon 1977), and migrants hailed from across the Southern African region.
[5] For example, Malkki 1992; Jing 2003; Turton 2005; Loizos 2008; James 2009.
[6] For example, Ferguson 1999; Gardner and Osella 2003; Parry 2003; Bakewell 2008.
[7] For example, Ferguson 1999; Ballard 2003; McKay 2003; Gamburd 2004; Mazzucato, Kabki, and Smith 2006.

in white-controlled centres of capitalist production like Johannesburg. It analysed how migrants lived their lives in labour compounds and in townships, in areas that were assumed to be 'white' spaces and where they had insecure tenure. And it investigated the relationship between migrants with different backgrounds while nonetheless concentrating – as did the migrants – on their long-term commitment to their more secure 'home' settings. With the contraction of formal sectors in Zimbabwe, South Africa, and elsewhere, and following Zimbabwe's political and economic crisis, 'displacement' – with its focus on upheaval and uprootedness – has replaced 'migration' as the dominant paradigm for understanding Zimbabweans' mobility.

However, these two paradigms – displacement and labour migration – need to be understood, not as alternatives, but in a dynamic relationship. Zimbabweans are dislocated. But on the border they find themselves drawn into a world in which white agricultural estates manage for the present to organise residents' lives, spatially, socially, and economically. Permanently employed farm workers – sometimes referred to simply as *mapermanent* – depict their lives on the farms in terms of stability and non-movement, asserting their rootedness as members of a resident population. Many such workers attempt to save for retirement, battling against Zimbabwe's economic and political instabilities in their bid to maintain their kin there. They establish stable structures of hierarchy and authority at Grootplaas that organise their own lives and the border compounds, and shape their relation to the border itself. It is these formalised labour hierarchies that seasonal workers are forced to negotiate. The latter, who come for the harvest, comprise both regulars and more recent recruits who seek work as they flee Zimbabwe's crisis.

To return to the Zimbabweans waiting at the Grootplaas gates: those admitted to the workshop yard, and into farm employment, *become* labour migrants *at their places of work*. Their relationships, their plans for the future, and their very sense of personhood come to be defined in relation to their workforce membership – with its paternalist hierarchies, its organisation of space, and its rhythms built around the wage.

What is more, traders are drawn by the lucrative markets represented by hundreds of waged workers. They hail from different national backgrounds (Zimbabwe and South Africa) and employment positions (farm labourers, displaced entrepreneurs, workers from other sectors, members of a former or aspirant middle class). Within and beyond the labour force itself, therefore, migrants with different motivations together make up a nexus point of provisional settlement and stability through the social arrangements on the farm.

Appreciating these experiences requires looking beyond displacement or neo-liberal fragmentation, and beyond a view of labour migration set apart from this fragmentation. Doing so, in turn, requires a second analytical move, which I have already prefigured. Rather than viewing contemporary phenomena in purely national terms, this book attempts to recover a regional view of crisis and employment.

For a Regional Perspective

Academic literature on Zimbabwe and South Africa has diverged markedly in recent years. This reflects the different histories of the two countries, but it underplays the political economy they have in common. Scholarship on the former, analysing the Zimbabwean crisis, has often focused on the politics surrounding the Mugabe regime.[8] Central in current analyses of South Africa, in contrast, has been a less personalised and more systemic analysis. Focusing on the diverse effects of neo-liberalism, this has covered themes from the shrinking of the state and outsourcing of its roles (Koelble and LiPuma 2005), to the rise of occult economies (Comaroff and Comaroff 1999), to the 'precarity' even of people with secure employment (Barchiesi 2011), to 'declarations of dependence' by insecure people seeking vertical ties of obligation and protection (Ferguson 2013). While Zimbabwe's and South Africa's forms of instability differ in many ways, what they share is their 'informalising' effects on ordinary people. Indeed, whereas Guyer (2004) writes of multiple 'formalisations' to characterise the diverse ways economic activities are brought within the purview of states, recent changes in Zimbabwe and South Africa might aptly be described as multiple but mutually reinforcing 'informalisations'. The shared experience of 'making do' in both countries is striking given apparent national contrasts.

The view from the border underlines the need for a regional historical perspective. When Zimbabweans cross the border into South Africa today, they encounter a rural landscape that bears the mark of similar settler-state policies to those that shaped Zimbabwe: sharp divisions between white commercial farms and black communal land. On white farms themselves, black workers live in labour compounds, on separate parts of estates from the gated houses of white farmers and managers. Farming ideals have promoted the figure of the fatherly farmer looking after his 'people', conferring gifts as he sees fit, rather than engaging in contractual arrangements with clearly defined limits (for South Africa, van Onselen 1992; see also du Toit 1993; Waldman 1996; for Zimbabwe, see Rutherford 2001; 2003). I describe regional white farming in greater detail in Chapter 2.

Zimbabwe and South Africa share wider historical legacies that are key to understanding the character of border farms. The two countries

---

[8] See Hammar and Raftopoulos 2003; Raftopoulos and Phimister 2004; Ranger 2004; Ndlovu-Gatsheni 2009; Raftopoulos 2009.

together formed a bloc of state-backed settler capitalism around which the political economy of the region was built. Central to this political economy were established patterns of labour migrancy, following the mineral discoveries of the late nineteenth century.[9] Moreover, it was gold on the Witwatersrand that led to settler expansion northwards into Zimbabwe, in search of further wealth (Phimister 1988).[10]

Inducing migration was partly achieved by limiting black access to cultivable land. Both countries saw the gradual creation of rural labour reserves, whose subsistence agriculture supplemented black wages, enabling white employers to offer lower pay (see Wolpe 1972).[11] Yet the racial division of land and patterns of migrancy were messier than they sound. Change was more complex and regionally diverse than the

---

[9] Gold was discovered on the Witwatersrand, the hilly area around present-day Johannesburg, in 1886. By 1900 there were already almost 100,000 black workers on the gold mines. Rapid change in South African economic structure and restructuring of the labour market shaped a system characterised by organised labour supply, oscillating black worker migration, the colour bar (with certain occupations reserved for whites), and industrial relations determined by strong white miner unions (Wilson 1972).

[10] The discovery of the Witwatersrand's gold underpinned 'the immediate genesis of colonial Zimbabwe' (Phimister 1988: 4). Mineral discoveries brought new importance to the southern African interior, and Cecil Rhodes hoped to find a 'Second Rand' north of the Limpopo River (ibid.: 5, 6). These mines, like those to the south, quickly became dependent on migrant black workers, secured by hut taxes and professionalised recruitment. In attempting to satisfy labour demands, they were brought into direct competition with their southern neighbours: 'the Rhodesian Native Labour Bureau was specifically designed to exclude the Rand from certain recruiting grounds and to direct labour to local mines away from those of the Transvaal' (ibid.: 50).

[11] The high levels of black labour migration in both Southern Rhodesia and the Transvaal were actively encouraged by state authorities. Although early migrants went to white centres of production to acquire items such as guns (see, for example, Delius 1980), moving to work was soon the result of deliberate government policies. A series of commissions in Southern Rhodesia allocated reserves for black Rhodesians, a process justified by reference to the idea that European civilisation was being imparted to eager Africans. But the majority of territory was kept for future white use. Meanwhile reserve land was of poor quality and was in the following decade further depleted by population growth (see Palmer 1977). In South Africa, it was the 1913 Natives Land Act that formalised racialised land ownership and access, allocating 8 per cent of land for black African occupation. This represented 'a degree of land alienation unrivalled in any sub-Saharan context' (Beinart 2001: 10). As in Southern Rhodesia, the Act rendered black South Africans' rural livelihoods unsustainable, and further reinforced oscillating labour migration patterns. Africans' wages as migrants supplemented inadequate agricultural yields on overcrowded land, and were used for the payment of taxes. One objective of racialised land allocation was to support white commercial agriculture, by supplying good land while squeezing out black competition. In South Africa, this 'undermin[ed] agricultural commodity production developed by Africans (often on white-owned land) during the previous half century', while causing a gradual shift to wage labour in increasingly capitalist white farming (Bernstein 1996: 5). In Southern Rhodesia, similarly, black producers became dependent on a range of commodities during years of agricultural prosperity, which then had to be satisfied through labour migration once their farming operations were squeezed in a process of 'delayed proletarianisation' (see Arrighi 1970; Kosmin 1977; Phimister 1988).

Introduction   11

classic Marxist 'transition' from 'feudal' to 'capitalist' relations (see, for example, Trapido 1978; Beinart et al. 1986; Bradford 1991; Beinart 2001).[12] One factor in this complexity was that white employers in rural areas – farms and provincial mines – sought workers from far-flung parts, often across national borders, because local populations – attracted by other forms of wage labour, particularly mining – were reluctant to work for them. The outcome of a century-long struggle between 'gold' and 'maize' (Trapido 1978: 53), over agricultural prices and labour supply (Bernstein 1996, 2004), was that farmers addressed their perennial 'labour problem' by using recruits from more peripheral areas of the regional political economy.

White farmers in both countries have depended constantly on workers from elsewhere – from *de facto* labour reserves of differing degrees. While agrarian capitalists in the Transvaal relied heavily on Rhodesian migrant labour (see Bradford 1993), farm workers in Southern Rhodesia hailed from Malawi and Mozambique (Rutherford 2001; 2003). The recent use of migrant workers in South African farming, 'to construct a cheap and manageable workforce' as a response to market liberalisation (Johnston 2007: 520), has longstanding precedence in the form of a floating reserve of labour, especially near border areas. Even in the Cape, where labour is considered to be 'local', farm-worker populations experienced the widespread fragmentation of families in a mobile population (see, for example, Waldman 1996).

The lives of many black southern Africans were therefore profoundly shaped by experiences of labour migration. Regarding Zimbabwe in particular, a recent survey showed that 39 per cent of a sample of black Zimbabweans had parents or grandparents who had been labour migrants outside the country, mainly to South Africa (Tevera and Chikanda 2009: 14). As early as the beginning of the twentieth century, 'a flow of migrant labour created within a *regional* economic system'[13] was a central feature of Southern Africa, and in particular of what would become Zimbabwe and South Africa. Black Rhodesians did not migrate to work only at centres of production in their own national state's territory; many headed southwards, hoping to attain better conditions and earn higher pay on the Witwatersrand (see van Onselen 1976). By the 1980s, Murray (1981) would conceptualise the region's migrant labour system in terms of a series of black rural areas, dependent to a greater or lesser extent

---

[12] In South Africa, the continuing access to land of black cultivators 'helps explain critical historical issues such as...the predominance of migrant labour' rather than full proletarianisation (Beinart 2001: 14–15). But land allocation also masked a number of non-capitalist arrangements between white landowners and black residents – tenancy, sharecropping, clientage.

[13] Mining Commissioner Bulawayo, 1904, in Phimister 1988: 55, my emphasis.

on white centres of production.[14] Chapter 3 considers how such regional mobility shaped the Zimbabwean–South African border, as it became a centre of capitalist production in its own right.

Unsurprisingly, labour migrancy became central to scholarly analyses. But this scholarship had its limitations. From the 1970s and 1980s, a neo-Marxist/revisionist political economy approach, related to an anti-apartheid stance in South Africa, lent itself to seeing divisions among the dominated as effects of capitalism, as preventing the growth of class consciousness, and as perpetuating an exploitative system. This framework had explanatory power, but was inadequate. In functionalist vein, it viewed state policies as expressions of capitalist interests (Bonner et al. 1993: 1), neglecting apartheid's complexity and contradictions (ibid.) and inferring strategy from consequences (for example Meillassoux 1981). Avoiding considering the experiences of migrants themselves, anthropologists sought the causes and effects of proletarianisation and exposed rural poverty and dependency (Hammond-Tooke 1997: 180). What it meant to leave home and be incorporated into workforces did gradually receive greater attention.[15] Yet, even with this broadening focus, concentration on labour migrancy privileged a particular view of the region, and risked masking other forms of regional mobility.

Assuming migrants have similar *class-based status* backgrounds led to the privileging of *ethnicity* as the key category of difference. Of primary concern was the construction and reconfiguration of ethnicity in mine compounds or towns, with especial focus on 'tribal' tensions and violence (see McNamara 1978; 1985; Moodie 1980; 1983; Moodie with Ndatshe 1994). This came at the cost of considering *other* sources of difference among migrants – McNamara's (1978; 1985) research in mine compounds suggests that clashes between Rhodesian and other miners were primarily about class-based status, not ethnicity *per se*. Divisions among workers were treated as immutable, structural outworkings of labour arrangements, rather than the products of workers' wider self-understandings.

---

[14] Specific agreements between governments officially enabled the recruitment agency, the Witwatersrand Native Labour Association (WNLA, known colloquially as Wenela), to recruit for the gold mines. Although they did not benefit from Wenela's recruitment efforts, many white commercial farms in the Transvaal and in Southern Rhodesia relied on this large-scale mobility.

[15] For some, this took the form of symbolic analysis, making sense of labour migrants' experiences and rituals through focus on 'cosmology': for example, seeing rituals of departure and return as a means of controlling migrants, a collective response to perceived dangers of urban/mining life (McAllister 1980; 1985; 1991; see also, for example, Comaroff 1985; Comaroff and Comaroff 1987; 1989; 1990). Others saw migrant experiences as more fragmentary, with individual agents pursuing strategies to make the best of their situations: for migrant mine workers, there were advantages to emphasising the very ethnic divisions that suited management (for example, Guy and Thabane 1991; see also 1987).

Introduction 13

More fundamentally, there has been a long history of labour migration from Zimbabwe to South Africa which does *not* fit the established migrancy framework. Even before the recent increase in numbers of Zimbabweans crossing the border, Zimbabwe–South Africa migration had been entrenched for over a century. However, its patterns and effects varied over time. As Chapter 3 of this book shows, crisis-related displacement often underlay Rhodesian labour migration. Peaks of Rhodesian migration to South Africa were the result of famine and violence at home, throughout the twentieth century. In western Zimbabwe during the violent anti-dissidence campaign of the 1980s known as *Gukurahundi*,[16] labour migrants 'were terrorised . . . as ex-combatants accused of going to South Africa to train to be "dissidents", by the police' (Werbner 1991: 157). For both former guerrillas and other able-bodied men, assumed to be dissidents, rural Matabeleland was dangerous. South Africa, Botswana, and Zimbabwean towns provided refuge. A year after the 1988 amnesty declaration, most young men, worried about returning, remained in South Africa as labour migrants. 'Their absence was due in good measure also to the impoverished state of the countryside, following the looting of crops, cattle and personal goods and, most recently, the run of bad years and severe drought' (ibid.: 162–3). In this area, labour migration was related to marginalisation by the Zimbabwean state, the result of persecution and perceived evidence of dissidence. It was especially associated with ZIPRA because many of its fighters in the liberation war had been labour migrants to South Africa (Bhebe and Ranger 1995b: 8). These examples of why Zimbabweans actually went to South Africa caution against overdetermining regional mobility with a single 'migrancy' logic.

Despite earlier over-emphasis on *labour* migration, as opposed to a broader view, this scholarship illuminated how migrants' experiences, strategies, and goals were shaped in adverse conditions. In this way, such literature has drawn attention to what everyday arrangements migrants produce at their workplaces. It has shown how black town dwellers historically struggled against the constraints imposed on them by the

---

[16] This targeted former Zimbabwean People's Revolutionary Army (ZIPRA) liberation fighters. ZIPRA and the Zimbabwean African National Liberation Army (ZANLA) were rival liberation armies, the armed wings of the Zimbabwean African People's Union (ZAPU) and the Zimbabwean African National Union (ZANU) parties respectively. ZANU emerged dominant post-independence. ZIPRA's recruitment area had been Matabeleland in western Zimbabwe, and it was considered still to be a ZIPRA stronghold. Its largely Ndebele population meant that ZIPRA was represented as an Ndebele movement – even though ZIPRA enjoyed little continued popular or ZAPU support. The region's population was targeted by the ZANLA-dominated Zimbabwean National Army. After this campaign, ZANU nominally incorporated ZAPU and changed its name to ZANU (Patriotic Front), declaring an amnesty in 1988 (see Alexander, McGregor, and Ranger 2000; Worby 1998; Werbner 1991).

migrant labour system, which assumed that their homes were elsewhere, in rural reserves. Despite insecure tenure rights, township dwellers asserted a sense of belonging by adapting their housing (Ginsberg 1996; Lee 2005). Some made use of formal employment by whites as bases for other enterprises, making the best of interstices in racialised systems of control (Bozzoli 1991; Bozzoli with Nkotsoe 1991; Preston-Whyte 1991). Others, those with middle-class aspirations in Zimbabwe, sought to define themselves in *opposition* to stereotypes of labour migrants (for example, Burke 1996; Scarnecchia 1999; Barnes 1999; West 2002). All of these efforts to establish and maintain dignified lives have their parallels on the border farms that are the subject of this book.

On the border farms, workplaces shape the experiences of their residents in ways that require analysis through the lens of labour migrancy. At the same time, the workforces reflect different, but mutually reinforcing, forms of instability. They are composed overwhelmingly of Zimbabweans who have undergone brutal upheaval at home. Workers also face uncertain futures, as white farmers seek to remain mobile to mitigate current risks in South African agriculture, apparently engaging in more flexible forms of capitalism. Workforces on the border today lie at the intersection of these processes, incorporating people, shaping the landscape around them, and mediating crises. What follows examines these crises in greater detail.

Zimbabwean Crises

Many of the farms themselves are products of cross-border migration of a kind often overlooked, in that several of the white border farmers hail from Zimbabwe. They came in the early 1980s to avoid black rule. Farming in Zimbabwe in the past, especially during the Liberation War,[17] shaped how they now think about farming across the border in South Africa. Border farmers who previously fought as counter-insurgents assess and evaluate their experiences in South Africa – the 1980s 'border war' and later farm attacks – with reference to this earlier period of their lives. Despite white farmers' mobile pasts, and uncertain futures, their enterprises along the Limpopo have become the workplaces and homes of thousands of black Zimbabweans, whose motivations for

---

[17] After white Rhodesia's Unilateral Declaration of Independence from Britain in 1965, ZANU and ZAPU began a fifteen-year armed struggle for majority rule of what would become Zimbabwe. The gradually intensifying conflict was concentrated in rural areas. White farmers, isolated on their land, led militarised lives characterised by periods of army service, armed transport convoys, and gatherings for target practice. For analyses of this period see Ranger 1982, 1985; Lan 1985; Bourdillon 1987; Kriger 1992; Bhebe and Ranger 1995a, 1996; Mtsi, Nyakudya, and Barnes 2009. For farmers' experiences of the war, see Godwin and Hancock 1993; Fuller 2002.

leaving Zimbabwe reflect that country's regional politics and recent history of successive upheavals.

The Production of a Cross-Border Mobility

In border farming operations, the ethnic and linguistic diversity of the workforce – often marked off in hierarchical terms – is the result of different experiences of crisis and reasons for mobility. In Limpopo, many permanent workers are from the far south of Zimbabwe. This is itself a diverse area, but the border's workforces are dominated by TshiVenda speakers, a small minority in both Zimbabwe and South Africa. It is their marginality that makes farm employment in South Africa so important to them. Zimbabwe's border region, in particular, has been sidelined in terms of infrastructure, education, land quality, and employment opportunities (Mate 2005). This area is in Matabeleland South, a province that experienced violent persecution in the 1980s. One strategy to avoid this violence was to cross the border into South Africa (see Werbner 1991).[18] Since the 1980s, when commercial farms established a demand for labour on the South African side of the border, Zimbabweans from just north of the Limpopo River have responded to hardship at home by working on these farms. They have established homes in the labour compounds on the border estates, adapting their houses to assert their rootedness despite the fact that they are sited on employers' land. And they have developed networks of kin, friends, and domestic relationships within and among the farm workforces. Border dwellers share TshiVenda as an everyday language. They moreover share self-understandings and aspirations, in part because of their predominantly rural upbringings and lack of formal education. Many have had long careers on the farms and their workplace seniority there is a source of status.

The 1990s saw Zimbabweans from increasingly diverse socio-economic backgrounds engaging in cross-border migration, as structural adjustment drove a rising number to seek work in South Africa (Zinyama 2000). The 1980s in Zimbabwe had seen high state spending on public goods such as education, fuelling the occupational aspirations

---

[18] The liberation movement ZAPU was recruited predominantly from Matabeleland, and after the election of ZANU to government at the end of the Liberation War, the continued existence of ZAPU dissidents was taken as justification for the massacre of more than 20,000 civilians by the ZANU (PF) government's Fifth Brigade (Muzondidya 2009: 179). There are also high levels of mobility further along Zimbabwe's southern border, further to the northwest within Matabeleland South. In a study conducted by Maphosa on Zimbabwe's border with Botswana, 68.7 per cent of sampled households had at least one member who had been a migrant in South Africa. Maphosa reports an unpublished survey by Hobane from 1996, in which 62 per cent of the working population in the same place worked in South Africa or Botswana (Maphosa 2005: 7).

of a generation of school students. However, 'the limitations of a welfarist programme unaccompanied by sustained economic growth soon became apparent' (Raftopoulos and Phimister 2004: 357). The economic restructuring that followed in the 1990s left many vulnerable, as increased responsibilities in formal sector jobs were paralleled by wage reductions, while other workers were simply retrenched. Zimbabwean migrants were both skilled[19] and unskilled, many of the latter seeking formal domestic work, agricultural employment, or informal trade.[20] For many, moving to South Africa in the 1990s to 'make do' was a sign that their earlier hopes of 'respectable' success had been shattered.

Increased cross-border migration was enabled by changes in South Africa itself. Following the end of apartheid and the relaxation of border controls, migration into South Africa from across the continent expanded significantly, although the magnitude of the increase is hard to measure and statistics are unreliable (ILO 1998: 8). Some Zimbabweans crossing to South Africa were recruited by the border farms, most of which were responding to the liberalisation of South African agricultural markets by turning to export crops. These crops, in turn, necessitated far larger workforces to maintain orchards and to pick, grade, and pack fruit.

The Post-2000 'Zimbabwean Crisis'

If the 1990s saw greater and more diverse cross-border migration, since 2000 the movement of Zimbabweans to South Africa has increased exponentially. The difference between the two periods is one of kind rather than simply of scale. This new trend was the result of hyperinflation, commodity supply crises, and political troubles at home, the main ingredients of what has come to be known as 'the Zimbabwean crisis'.

The causes of the Zimbabwean crisis are complex. The country's economy began to break down in the late 1990s under the continuing effects of structural adjustment, expensive military intervention in the Congo (see Raftopoulos 2009), and unbudgeted pay-offs to war veterans demanding recognition for their participation in the Liberation War (see Muzondidya 2009). Fiscal difficulties were compounded by a crisis of legitimacy for the ruling party, the Zimbabwean African National Union (Patriotic Front) – ZANU (PF) – when the party lost a nationwide referendum on constitutional reform in 2000. Having proposed an extended presidential term for Robert Mugabe, the new draft constitution was

---

[19] The Zimbabwe High Commission in Pretoria estimated 60,000 migrants in professional positions in the late 1990s (Fultz with Pieris 1997: Appendix II).

[20] Informal trade was an option Zimbabweans had been taking up since the mid-1980s, notably selling crotchetware and, in the 1980s, supplying spare motor vehicle parts and small machinery to commerce and industry. Such cross-border trade was increasingly 'dominated by women seeking to supplement their family incomes to clothe and educate their children' (Zinyama 2000: 73).

defeated by a coalition of churches, NGOs, and unions (the National Constitutional Assembly), which became the opposition party Movement for Democratic Change (MDC). This defeat marked the beginning of a more authoritarian style of rule characterised by horrific atrocities (see Dorman 2003).

Popular grievance had been widespread earlier in the 1990s, because structural adjustment had tended to benefit elites. Consequently, 'a combination of a slow-down in the state's land-reform programme, intensified pressure on land in the [black] communal areas, and economic liberalisation' in the 1990s led to a number of popular occupations of white farmland (Raftopoulos 2009: 211). As ZANU (PF) sought a support base in the face of newly vocal opposition, it provided state backing for these occupations: 'the occupations of 2000 were spearheaded by war veterans and actively encouraged by the President in a process that directly undercut the developmental state' (Alexander 2003: 113). This process culminated in a large-scale confiscation of land from white farmers known as 'fast track' resettlement. In the short term, the consequent destruction of the commercial agricultural sector sent the Zimbabwean economy into a further tailspin. But transfers of land had farther-reaching and more complex effects. Land redistribution – both in the form of A1 (smallholdings) and A2 (small- and medium-sized commercial farms) models – has represented a complete restructuring of the agrarian economy. It has meant the end of a broad divide between large, generally white-owned commercial farms and communally held land in black reserves. And the allocated land has presented new livelihood possibilities to many Zimbabweans beyond a narrow elite (see Scoones et al. 2010). Nevertheless, fast-track resettlement schemes are beset by problems: a lack of security of tenure, with many occupants holding at best only an official offer letter; the sheer speed of implementation, making it difficult for officials to respond to unexpected complications; ambiguity about who holds ultimate on-the-ground authority; low productivity, in the context of wider crisis and extremely limited government support (see Matondi 2012; also Cliffe et al. 2011). One result has been that many Zimbabweans who acquired new land nevertheless rely heavily on remittances from relatives abroad (Matondi 2012: 242).

More generally, the growing crisis since 2000 has led a huge number of Zimbabweans to seek ways of making ends meet outside the country. Most seasonal workers on South Africa's border farms arrived during this post-2000 period. The politics underpinning the crisis belie the sharp distinction between 'economic' and 'political' motives for migration.[21] In South Africa, 'the legal status options available to [Zimbabweans] – refugees fleeing persecution or economic migrants

---

[21] This is a distinction drawn by both South African and British authorities to deny Zimbabweans asylum (Kriger 2010; Engelke, personal communication).

trying to better their lives – leave most as "illegal foreigners'" (Kriger 2010: 81).[22] In reality, many people were driven to move by a complex combination of economic collapse, new forms of political exclusion, and direct coercion in Zimbabwe. ZANU (PF) has attempted to redefine Zimbabwean citizenship: whites, farm workers, urban workers 'without totems',[23] women, and members of the political opposition have become excluded from a nationalist project cast as a continued liberation struggle (Raftopoulos 2003; see also Hammar 2003; Ranger 2004).

Some recent Zimbabwean migrants to Limpopo farms, disproportionately seasonal workers, are precisely those disqualified from full citizenship by the Zimbabwean state, including MDC activists who have been victims of Zimbabwean state repression. Those with urban backgrounds often have few connections with the countryside, and lack the legitimation of totemic clan emblems. Some of these left Zimbabwe following May 2005's Operation *Murambatsvina* ('the one who rejects [clears away] the rubbish'): a vast state programme to demolish 'illegal' housing and informal businesses, and send occupants 'back' to familial rural areas to which often they had little or no connection (see Potts 2006). A month later, the clearances had 'resulted in an estimated 300,000 displacements of civilians in urban areas countrywide, with mass loss of livelihoods and property' (SPT 2005). By September, '2.9 million people across Zimbabwe were in need of food aid', while inflation reached 359.8 per cent and unemployment 80 per cent (HRW 2006b: no page).[24]

---

[22] In July 2008, a refugee reception centre was opened in Musina by the Department of Home Affairs, offering renewable asylum-seeker permits. But, while asylum-seeker permits offered a provisional way to remain in South Africa, and suggested official acknowledgement of the political reasons for migration, few asylum claims were actually approved. March 2009 saw the Minister of Home Affairs grant Zimbabweans a special visa status and order a suspension of deportations. Then, in 2010, the Zimbabwe Documentation Project (ZDP) required Zimbabweans who were not recognised as refugees to be regularised, using proof of occupation. Around 276,000 Zimbabweans were registered before the ZDP closed in 2011, making way for the resumption of aggressive deportations in late 2011 (Derman 2013; IRIN 2013; on changing border policing, see Rutherford 2011).

[23] Suggesting a lack of proper attachment to rural clans, and therefore a lack of deep autochthonous belonging.

[24] This 'slum clearance' has been seen as part anti-opposition measure and part commitment to modernist 'planning' (Potts 2006). At heart, however, it was a starkly violent assertion of ZANU (PF) power. In July 2005, a United Nations report on *Murambatsvina* stated:

> It is estimated that some 700,000 people in cities across the country have lost either their homes, their source of livelihood or both. Indirectly, a further 2.4 million people have been affected in varying degrees. Hundreds of thousands of women, men and children were made homeless, without access to food, water and sanitation, or health care. Education for thousands of school age children has been disrupted. Many of the sick, including those with HIV and AIDS, no longer have access to care. The vast majority of those directly and indirectly affected are the poor and disadvantaged segments of the population. (Tibaijuka 2005: 7)

Other residents of the South African border farms, until the land invasions, had been experienced farm workers in Zimbabwe, where they had long been stigmatised because of their outsider, and quasi-feudal, status. Many of Malawian or Mozambican descent, they were seen in an ambiguous light – oppressed by, but simultaneously people belonging to, the farmer (Rutherford 2001; 2003). 'Land reform' for them meant deliberate brutality rivalling the urban experience just described.[25]

The deepening crisis, however, meant that many recent Zimbabwean migrants to South Africa fit none of these categories, but are drawn from skilled, educated, even middle-class backgrounds. During the fieldwork on which this book draws (2006–8), hyperinflation rendered many formal occupations inadequate, making them less viable than minimum-wage work paid in rands on South African farms.[26] For some, it was not that they lost their jobs, but rather that their pay simply no longer covered transport to and from work. Others came of age as employment possibilities contracted. Further, the Zimbabwean government responded to rising consumer costs by fixing commodity prices by decree. The result was that retailers, rather than sell at a loss, stopped stocking controlled goods. South Africa's attraction was that it offered not only the prospect of pay in rands, but also access to key supplies no longer procurable in Zimbabwe.

Such economic deprivations always have an explicitly political side: those who could not find or retain government employment were often those who refused or failed to enlist in youth camps and become Green Bombers (youth militia; Rutherford 2006). The political causes of Zimbabwe's crisis are clear to labourers on the border farms. Workforces are vocally anti-ZANU (PF), something especially noticeable during the 2008 elections when many wore MDC T-shirts. The elections were bitterly contested. Morgan Tsvangirai's faction of the MDC won the parliamentary election. But, despite Tsvangirai's lead, the presidential race was officially declared inconclusive. The run-off election further

---

[25] A study by the General Agricultural and Plantation Workers' Union of Zimbabwe (GAPWUZ) described it as:

> A violent, state-sponsored and systematic attack on 1.8 million people in order to wipe out any illusions of political freedom they might have cherished, to force them into the ranks of strict ZANU-PF orthodoxy and to prevent them from lending support to the fledgling Movement for Democratic Change (MDC) opposition party. (GAPWUZ 2010: 7)

The violations reported by GAPWUZ's farm-worker respondents included assault (68 per cent of respondents), torture (66 per cent), children forced to watch beatings (38 per cent), rape (11 per cent), and murder (10 per cent) (ibid.: 33).

[26] Minimum wage R989 during fieldwork. Of seasonal employees, packshed workers earned the most at around R1,000 because of overtime. Pickers received in the mid-R800s, because they were paid a piece rate calculated from the hourly minimum wage by assuming an overly ambitious work pace. In 2007, the exchange rate was around one British pound to R14.

underlined the ZANU (PF) regime's responsibility for cross-border displacement, when flight from state-sponsored reprisals against opposition voters brought recruits to the farms. Perpetrated by party supporters, war veterans, and security services, the wave of violence included harassment, intimidation, beatings, rape, abduction, murder, and the burning of homes and businesses (see, for example, IRIN 2008).

Under the Government of National Unity that followed the 2008 elections, Tsvangirai became Prime Minister while Mugabe remained President. Yet ZANU (PF) retained control of the police and the army. Hyperinflation ended in 2009, when foreign currencies including the US dollar and the South African rand became legal tender and the Zimbabwean dollar was soon abandoned. Nevertheless, Zimbabwe's plight lingered on, 'most dramatically summed up by the Human Development Index for 2010 which placed Zimbabwe last in the world' (Derman and Kaarhus 2013: 8). Even with subsequent improvement, Zimbabwe's economy remains unable to accommodate most people, and unemployment was estimated at over 80 per cent in 2013 (IRIN 2013). Many Zimbabweans remain dependent on earnings in South Africa, including on the border farms just across the Limpopo.

Recent migrants, in sum, have a wide range of socio-economic backgrounds, from former manual workers to former teachers with qualifications. They have come to South Africa's border farms at different times, responding to different, interrelated pressures, both economic and political. And they vary, in turn, in their patterns of mobility and settlement.

How have the experiences of recent Zimbabwean migrants been analysed by scholars? Zimbabwe's 'new diasporic communities formed since 2000' have encountered 'exclusion, distress and hardship' (McGregor 2010: 27). Considering the Zimbabwean case, Mbembe interprets displacement as experience of 'radical loss', of 'consciousness of the precariousness of life, and of the lack of guarantees to its future' (unpublished conference keynotes quoted in Hammar, McGregor, and Landau 2010: 267). Stepputat similarly emphasises a 'phenomenology of loss', and Guyer a sense of 'dislodgement' – disruption that can occur even without spatial movement (unpublished conference keynotes quoted in ibid.: 268). Worby contends that, for displaced Zimbabweans, 'the temporal horizon for reconciling credits and debts, present capacities and future dependencies, is unknowable and thus in some profound way, undecidable' (2010: 421). The consequent sense of dramatic 'abjection' (Ferguson 1999) is particularly acute precisely because of Southern Africa's history of highly structured economies:

Displacement has deepened informalisation. While this has parallels in other African settings, the process is unique in Southern Africa because of its relatively well developed formal economies, and because of the way informalisation articulates with residues of older modes of bureaucratic power in a region with histories of particularly strong states. (Hammar, McGregor, and Landau 2010: 282)

Displacement, in short, equates to deeply disruptive, forced mobility, which also foregrounds flux and innovation. The flip side is that 'new spaces, agents and dynamics of exchange and accumulation' are being created by it (ibid.). *But what new spaces?* 'The boundaries between displacement and migration' need rethinking (ibid.). On the South African border, Hall (2013) identifies different categories of migrants to the farms: 'first-time migrants in distress', 'long-term oscillating migrant workers', 'passing-through migrants', 'short-term savers', and those who come to the farms for the purposes of targeted theft. These are all in evidence at Grootplaas. Hall's typology helpfully illuminates migrants' senses of their own options in the context of crisis. It opens up analytical conceptions of mobility on the border. Yet plans change, as Zimbabweans respond in different ways to the farm setting. Even those who stay in the labour compounds without employment are drawn into wider patterns of migrancy. They become part of labouring populations, organised according to steep hierarchies and established work processes.

Zimbabweans' contemporary mobility cannot be understood without appreciating Zimbabwe's particular history of successive upheavals, and analysis must avoid the depoliticising effects of seeing them *only* as labour migrants (see Rutherford 2010: 72). Nevertheless, migration to the farms on South Africa's northern border is a specific instance of a much wider, historically established pattern of migrancy and incorporation into waged workforces.

There is another way in which participation in the farms' workforces ought not be seen simply as the result of an uprooting from Zimbabwe. This is a regional not a national story. As I have noted, the farms are affected by their wider context on both sides of the border, characterised by different kinds of instability. If one is produced by the Zimbabwean crisis, the other is often glossed as the result of 'neo-liberalism'.

Markets and Mobile Agriculture in Neo-Liberal South Africa

Possibly the most widely noted source of current instability, in the region and elsewhere, is globalised neo-liberalism, infused with market-driven short-termism. Despite the 1994 ANC-led government's radical mandate, it soon introduced austere neo-liberal monetary policies to South Africa (Bond 2000). Neo-liberalism has been held up as the cause of diverse phenomena, from renewed expressions of autochthony (Geschiere and Nyamnjoh 2000) to the resurgence of chieftaincy and tradition in South Africa's 'African Renaissance'[27] (Oomen 2005). Extremely diverse socio-economic phenomena can be understood as 'concrete, historically specific outworkings of millennial capitalism and the culture of neo-liberalism' (Comaroff and Comaroff 2000: 334). Migration has become an expression of global capitalism: 'that the new and ever surging waves of migration are linked to the accelerated globalisation of consumer capital is all too obvious', notes Nyamnjoh (2006: 31).

Pro-market reforms have been a key source of post-apartheid disillusionment. Many South Africans hoped the end of apartheid would bring new political inclusiveness and socio-economic justice. Such expectations have not been fulfilled. What change there has been is often believed to have benefited a small elite. 'The poorest 50 per cent of the population are worse off economically than they were under apartheid' (Robins 2005: 2). Neo-liberal policies of privatisation and state-downsizing have resulted in the outsourcing of rural governance to 'traditional' authorities, in a manner reminiscent of apartheid. (Koelble and LiPuma 2005).[28]

At the heart of all this is the fact that the South African transition from 'apartheid' to 'post-apartheid', 'repression' to 'freedom', has been accompanied by and associated with others: from 'state' to 'market', from 'racial Fordism' to 'non-racial post-Fordism' and from 'rigidity' to 'flexibility' (Hart 2002: 25). South African farmers are faced with the uncertainties and pressures of such a world. In crop agriculture, the rule tends to be: 'get big or get out'. Among those who survive, aeroplanes, tennis courts, and plush houses attest to their success. Most border farmers switched to export fruits from cotton in the 1990s, after agricultural market liberalisation[29] and the import of cheap textiles from

---

[27] A term popularised by South Africa's president Thabo Mbeki, envisioning a future both of 'modernity' and, more relevant here, one 'in which African communities succeed in constructing themselves around tradition, legacy and heritage' (Lodge 2003: 230).

[28] One consequence has been widespread xenophobia. In much of South Africa, foreigners, known as *makwerekwere* in imitation of their incompetence in South African languages (Nyamnjoh 2006: 39), are blamed for a variety of social ills (see, for example, McNeill 2009). Strong anti-foreigner sentiments are reinforced by the reimagination of citizenship in South Africa. During apartheid, black South Africans were conscious of sharing a predicament – oppression and exploitation at the hands of an Afrikaner Nationalist regime and white capital – with black non-South Africans. But after 1994, this *regional* inclusivity was compromised by efforts to establish *national* unity (Jensen and Buur 2007). It is notable in passing that, on the remote border farms, the increasingly vitriolic xenophobia to the south, especially towards black Africans, appears distant. Farm residents hear about incidents – such as the outbreaks of violent xenophobia in Johannesburg and Cape Town in 2008 (see Morreira 2010) – on television or from relatives, rather than experience them directly.

[29] The huge sums of aid that the South African government had given to white farmers since 1948 had already become too expensive by the 1980s and were cut. But in the 1982 drought 25,000 farmers still received 'R2.7 billion in loans, conversion grants and drought relief' (Lodge 2003: 71). The later 1980s saw the doubling of debt in agriculture (ibid.). One result of these changes was the consolidation of farms into bigger operations.

China. Grootplaas moved into citrus – oranges and grapefruits – sent via buying agents to destinations across Europe and Asia. They joined other southern-hemisphere growers who, by filling the seasonal gap, ensure oranges' year-round availability as the world's 'first fruit crop in international trade in terms of value'.[30] Grootplaas's last ship consignments, which arrive in Europe around six weeks after fruit is picked, need to clear customs before the end of an import window, after which tariffs climb steeply. This tight schedule reflects the fact that the border farms' survival depends fundamentally on sales in an international market economy.

Given the integration of South African farms into international markets, it is certainly the case that changes in agriculture are in part an effect of global economic instability and casualisation. Since the 1970s, Fordist labour arrangements have given way to 'flexible accumulation': 'the more flexible notion of capital emphasises the new, the fleeting, the ephemeral, the fugitive and the contingent in modern life, rather than the more solid values implanted under Fordism' (Harvey 1990: 171). 'Global' restructuring of work organisation to ensure cheap, flexible labour has in turn generated the growth of the informal or 'underground' economy (see Castells and Henderson 1987; Sassen-Koob 1987), and led skilled workers to be redefined as unskilled to cut costs (see Blum 2000 for San Francisco shipyards). South Africa has seen

the steady withdrawal of capital from fixed, large-scale, labour-intensive production, resulting in extraordinarily high rates of unemployment. This is transforming the nature of change itself in a society where industrialisation was long held to be the motor of historical development. (White 2012: 399)

Although analysts of such trends have tended to have factories or urban corporate capitalism in mind (Ortiz 2002), a similar story applies to agriculture. In South African farming, there has been a move towards a 'leaner', suppler model. White farmers today face a liberal buyer's market. In the 1990s, protective marketing boards were largely dismantled and subsidies discontinued. Farmers avoid attachment to their enterprises, and are keen to remain flexible in the face of an uncertain future. Agriculture starts to be viewed as a series of strategic business investments, rather than as a necessarily rooted way of life. This involves seeking new opportunities to develop land, stay mobile, and distribute risk across different enterprises and crops in different regions. It also involves 'the intensification of the *fragmentation of labour*' in which potential workers 'pursue their reproduction in conditions of increasingly scarce, insecure

---

But the new dispensation after 1994 went further, leaving commercial agriculture to the mercies of the open market.

[30] <http://www.unctad.org/infocomm/anglais/orange/market.htm> (accessed 12 April 2010).

and oppressive wage employment' (Bernstein 2007: 45; original author's emphasis). On the Zimbabwean border, this is further enabled by the particularly high degree of mobility in the area and the continual presence of military and police border patrols that keep new, undocumented recruits vulnerable. A large, transient labour surplus is a malleable and cheap solution, with workers easily controlled because of their 'grey' legal status.

It is not enough, however, to view all of this simply as an instance of neo-liberalism. Labour arrangements on the border, which enable the existence of an apparently more 'globalised' industry, are nonetheless the result of particular, diverse political phenomena, rather than simply of generic 'global' processes. They arise out of South African attempts to redress apartheid injustices, and out of Zimbabwe's political and economic crisis. Foremost among the uncertainties that farmers face are the unpredictable vagaries of post-apartheid land reform (see James 2007). As employers, they attempt to cultivate an apparently neo-liberal, corporate image precisely to create distance from the paternalist style that characterised earlier agricultural arrangements – now highly symbolic as an unwanted anachronism. They also de-emphasise their discretionary responsibilities to workers in reaction to state requirements of a minimum wage and working conditions. Changes are *political* economic, rather than simply attributable to any abstract notion of 'the market'. Analyses of globalisation often risk collapsing diverse histories into flattened accounts (Cooper 2005). What appears merely the result of global forces may equally reflect regional or national processes. All such processes, moreover, are manifested through actual arrangements in particular workforces. Flexible capitalism here emerges from farmers' responses to diverse pressures – pressures that cannot simply be glossed as 'neo-liberal'. Neo-liberal effects here are partially divorced from neo-liberal causes. Indeed, the result can perhaps best be described as 'accidental neo-liberalism'.[31]

Even such effects cannot be taken at face value. Neo-liberal versatility often relies on established, even rigid, forms of organisation on the ground. As Harvey argues, the advantages to capitalists of changing location or product in pursuit of profit are often offset by the costs of investing in new physical infrastructure and immobile inputs (1990: 234). Such costs contribute to maintaining a status quo. On farms, for example, the flexibility allowed by uprooting one's operation and transplanting it in a more stable, less threatening political environment is counterweighed by the need to invest – in however short a term – in expensive infrastructure. Packsheds are extremely costly, especially when mechanised. Citrus trees take several years to produce saleable fruit, after which they continue producing for around 30 years. Although a period of upheaval is now

---

[31] My thanks to Keith Breckenridge for suggesting this term.

occurring in South African agriculture, matters are far more complicated than might be conveyed by an image of capitalists darting around the globe, in hypermobile pursuit of profit. Success in any one farming enterprise requires the investment of considerable time. Understanding current South African economic upheaval therefore requires attention to the fact that flexible accumulation is itself dependent on highly non-flexible and structured arrangements.

If neo-liberal mobility has its limits, surely the Zimbabwean crisis nevertheless plays right into the hands of farmers as potentially flexible accumulators? The destitute, vulnerable people who come through the border fence do indeed represent a pliable supply of labour. However, the labour force itself also makes an agricultural enterprise more sedentarist and less flexible. Harvey (1990) makes the point that flows of capital are becoming ever faster, in what he calls 'time-space compression'. But in agriculture, capital flows are not always as transient as they may seem. Border farming on the Limpopo River crucially relies on the complex, highly personal hierarchies and organisation that emerge among workers. Forty years of border farming is not a long history, but it is long enough for hundreds of workers and their dependants to develop homes and lives on the estates. Current risks have indeed led farm owners to emphasise their potential flexibility rather than stressing the rootedness in place that formerly characterised – and may still in some cases be a feature of – white farmers' self-understandings. But there are limits to this in everyday operations. Such farms continue to rely on core black workforces of permanently employed, resident 'general workers'. They maintain all aspects of the estates throughout the year, and situate themselves as figures of authority and protection on the farms, producing what I call 'mediated paternalism' (Bolt 2013). An ethnographic study of one farm reveals the structured arrangements on which apparently fluid, neo-liberal capitalism depends.

Although the relevance of labour migration scholarship appears to be receding, its legacies remain crucial to understanding current realities in South Africa. Many people are forced to 'diversify their forms, and spaces, of employment (and self-employment) to meet their simple reproduction needs as labour ("survival")' (Bernstein 2007: 39). While a minority 'earn a living' through stable or insecure employment, most resort to 'making a living' more broadly (von Holdt and Webster 2005). Meanwhile, even workers with permanent jobs experience precarity – despite the place of regular employment at the heart of official discourses and popular aspirations of social citizenship (Barchiesi 2011). In a region where jobs have become scarcer, it is imperative to understand how different kinds of livelihoods, and different kinds of stability and instability, intersect.

Key here is how centres of employment reflect and shape their local surroundings, including the character and domestic arrangements of the labour force. On the border farms, core workers, casual workers, and informal traders form hierarchies that are not simply imposed by employers. The formal and informal economies constitute one another in important ways. Informal businesspeople rely on residence in labour compounds, and in turn offer workers the credit that gets them through the month. For some people, formal employment is even a means to become an informal trader and seek new opportunities for business in the compounds (see Chapter 7).

What characterised the region in the past was, precisely, the way it offered a precarious foothold in formal employment. The gold mines themselves, the paradigmatic case, illustrate the current relevance of labour migrancy scholarship: 'a volatile gold price results in the ever-present threat of retrenchments. Such insecurities prevent many mine-workers from living with their families in one household, as another home-base may be needed when jobs are lost' (Rabe 2006: 86). New forms of insecurity here take remarkably familiar forms, as miners are forced back into oscillating migrancy. Familiar patterns of labour hierarchy and labour migration have new significance. We now need to make sense of islands of waged workers amidst transience. Comprehending the ways workforces incorporate people, at the intersection of different historically established migratory logics, is key in this stable-yet-unstable setting.

Employment as Incorporation

Employment brings more than an income. It incorporates people, albeit often in steeply asymmetrical relationships. James Ferguson (2013) has argued that Southern African history over the last 150 years can be characterised precisely by people's attempts to find security in subjection. People subordinated themselves to different authorities – the nineteenth-century Ngoni state, employers on the mines and the farms in the twentieth century, and the post-apartheid state. Through the new social relations and forms of personhood that developed, they opened up new options in life. The Ngoni state and the mines were equally hungry for people: wealth lay in people themselves, or in their labour. The major watershed in Southern African history was therefore not the rise of white regimes, or their demise. It was the shift from a region hungry for people, to a region with a human surplus. This is a useful starting point for thinking about contemporary centres of wage labour. In an era of labour surplus, it is imperative to investigate how people gain a foothold.

But we need to go further than simply identifying a logic of incorporation. How, exactly, does incorporation change people's options? What forms of inequality are created, reinforced, or challenged through this incorporation – not just between employers and employees, but among workers? Who is excluded, and who declines integration altogether (self-exclusion)? This book offers the view from within a labour force, as people attempt to achieve provisional security.

Amidst the instability of today's mass unemployment and *ad hoc* strategies of making do, even highly unequal labour arrangements suggest possibilities. Being incorporated into a workforce alters the very terms of engagement with time and space. Countermanding the flux and everyday insecurity of the border area is the structure, regularisation, and provisional predictability of the farm's working day. Diverse personal histories and experiences of chaos and displacement are transformed into routine, even monotony, as residents become part of the labour hierarchy. Living from one day to the next (see Day, Papataxiarchis, and Stewart 1998) gives way to possibilities for planning. People's responses differ. Remitting to kin in Zimbabwe is weighed against investing in relationships in the compound, however short-lived. Low wages and the costs of living at the farm can make saving anything difficult. During the period of hyperinflation and the peak of the Zimbabwean crisis, remittance could be in kind, transforming workers into traders. But regardless of this diversity, as people join this waged workforce in a transient setting, the timeframes of goals and expectations extend.

To those who are recruited, Grootplaas becomes less and less like what Augé (1995) calls a 'non-place', devoid of the kind of social significance that invites attachment. Forms of workplace camaraderie connect strangers to each other, and to work supervisors whose roles extend beyond the orchard, field, or packshed. In places of capitalist agricultural production like Grootplaas – and indeed in Southern Africa's hubs of resident labour more generally – work entails relatively stable living arrangements, and chains of command at work mean round-the-clock authority. Marula, the senior foreman, is a headman-like figure. His work overseers act as elders alongside him and comprise his court. Senior workers establish themselves by building up retinues of dependants for whom they speak in disputes, and for whom they safeguard earnings. New recruits quickly learn the relationships and the forms of self-presentation that constitute the farm hierarchy.

But this is far from simple, or indeed uniform. Incorporation is often fraught, uncertain, and incomplete. People become rooted to very different degrees. And this complexity results from inequalities that are maintained within the workforce, not simply between employer and employees. A new recruit is certainly presented with new possibilities, but as he or she is shown to a bare floor in a dirty room in the rowdiest part of the labour compound, distinctions between workers are marked out. Because of overcrowding, seasonal workers spend most of their waking hours outside. At the height of the Zimbabwean crisis, they were often the targets of aggressive police deportation raids that swept through the border's compounds.

Moreover, actually achieving a degree of stability on the border means establishing connections to senior figures – connections that are often fragile. Hierarchies are highly gendered. The vast majority of Grootplaas's 140 *mapermanent* are men. The exceptions are a small number of domestic workers in the farm offices and white houses, and two shopkeepers in the compound store. Around twenty further women are 'semi-permanent' – well integrated in border farm networks, they are allowed to reside in the compound throughout the year, and are kept on call for piece jobs such as cleaning the compound or standing guard against baboons in the orchards. Hierarchical relationships on the border farms also intersect with experiences of Zimbabwe's recent crisis. For male seasonal workers with middle-class backgrounds or aspirations, the rough camaraderie involved means compromising norms of behaviour that connect them to their past lives – actual or imagined.

On the other hand, ties offer access to the quieter and more secure parts of the compound, where permanent workers have adapted accommodation. They bring the possibility of leaving wages with work supervisors, whose locked rooms mitigate the risks of living among mobile strangers (see Bolt 2014). More generally, good relations with senior figures increase the likelihood of receiving a work permit, and integrate recruits into the sociality of drinking groups. They offer the stable residential base – at least for a while – to establish informal businesses, such as selling cigarettes, snacks, or marijuana to other workers. Occasionally, seasonal workers who are felt especially to belong, or who are well suited to particular kinds of work, are brought into the permanent workforce – this, despite extremely low labour turnover among the permanently employed.

For the few who become permanent workers, rooting themselves on the border becomes considerably easier. *Mapermanent* have their own rooms, which they gradually adapt, transforming concrete-lined cells with furniture built from scrap in the workshop yard, and with *stoeps* (porches – here, extended front steps, sometimes with low walls) built from spare bricks and concrete. They also have work permits and stable incomes, and some acquire South African identity cards. All of this, in turn, establishes the conditions for a degree of congenial domesticity, as Chapter 4 reveals. Male permanent workers develop live-in relationships, and even longer-term 'farm marriages', with women who come to the farm to find work – either for a formal wage, or looking after children in the compound. For women in this situation, such relationships represent a form of incorporation into the relative stability of farm life. After all, male permanent workers are something of a labour aristocracy here.

This can be a double-edged sword, however. A few women gain a firm foothold. But far more, despite being sole providers for their immediate

Introduction

kin at home, experience impermanence and precariousness. The compounds' rules are overwhelmingly established and enforced by permanently employed men, who often have less invested in *particular* relationships than they do in having someone around. Permanent workers patch together the conditions of their domesticity from the transience around them. Their own everyday rootedness, in fact, reproduces the insecurity of mobile women: some are left with children, and an ever-greater dependence on future, fragile integration into farm arrangements. Hubs of wage labour along the border gather people and shape their options and their movements, even as they leave some far more vulnerable than others.

Workplaces are usefully seen as crucibles of wider processes, as anthropologists of the Manchester School showed. But this means starting by taking labour settings on their own terms, not simply importing categories from analyses of life outside them. Just as, for one such anthropologist, Max Gluckman (1961), an African miner has to be considered as a miner, not just a displaced 'tribesman', Zimbabweans on the border farms require understanding as farm workers. But this 'situational analysis' comes to mean something different amidst current, far-reaching changes in Zimbabwe and South Africa. Firstly, at Grootplaas, diverse experiences of displacement intersect with one another in workplace micro-politics and status hierarchies, in unpredictable ways. Recruits with middle-class aspirations eschew the performances of camaraderie through which the majority embed themselves in farm life (see Chapter 5). Rivalry between different kinds of work authority – foreman and personnel manager, field overseer and office clerk – intersect with rivalries between established farm-worker families from the border and people from further north with contrasting backgrounds and self-understandings (Chapter 6).

Secondly, a situational focus on workplaces requires attention not only to workers, as in classic labour studies, but also to both workplace histories and employers themselves. Enriched with different perspectives – employers and employees, past and present – this text reveals how what looks like neo-liberal flexibility is in fact underpinned by farmers' life projects, structured hierarchies, and complex everyday relationships among members of the labour force, all of which bear the imprint of a regional past.

Becoming a Grootplaas Resident

The research on which this book draws was conducted between December 2006 and April 2008, a time of economic freefall and the run-up to important elections north of the border. For 17 months, I lived in Grootplaas's labour compound. During the harvest, I also worked as a picker in a 30-man gang.

As a white, British anthropologist, my residence in a farm's labour compound was unprecedented, and requires some explanation. I had visited Grootplaas during a two-week pilot trip to South Africa earlier in 2006, when I had been put in touch with Willem, the farmer. The timing of my return to commence fieldwork was serendipitous. It coincided both with a payday and with the HIV/AIDS peer-educators' Christmas party in the compound, where I spent the weekend, meeting people at the shebeens,[32] and later took up temporary residence on the floor of the two-room house occupied by Michael, the personnel manager. Soon after, I acquired my own room in the compound through the workers' Housing Committee, while continuing to eat, wash, and spend time at Michael's house. Although sanction for my presence on the farm was given by the white farmers, such endorsement came after arrangements had already been made with the workers with whom I lived.

Receiving permission from the farmers to live in the Grootplaas compound was undeniably a stroke of luck. Historically, farmers were known for being suspicious of researchers. In South Africa this followed 'exposé' reporting that portrayed Afrikaners in a very bad light (for example, Crapanzano 1985). In Zimbabwe, Rutherford (2001) was flatly refused when he asked Zimbabwean farmers if he could live in their compound. My access, in contrast, was in part a product of the extremely close monitoring – by supermarkets, agents, and government officials – to which such farmers were subject. A mere anthropologist was perhaps thought less threatening than these monitoring agencies to farmers' enterprises.

As I established myself in the compound, I encountered surprise from its residents, the farm's black working population. I made a particular effort to explain that my purpose in being there was to try to understand people's experiences of life on the farms (*vanhu vanorarama sei pamapurasi* – ChiShona: how people live on the farms) as an anthropologist. Long after I arrived, a few people, observing the time I spent writing fieldnotes at the compound's adult literacy centre each morning, continued to believe I was a computer technician. But several residents, having been formally educated, understood my purpose and explained it to others. The disconnect within the labour force – between exiled Zimbabweans of relatively high status and residents who had never been to school – sometimes took on bizarre forms. For example, the farm's storeman, Benjamin, educated to A-level standard, discussed with my PhD supervisor the politics behind recent professorial appointments in Zimbabwe and South Africa as we walked around the crude one-room houses in the labour compound. On the same visit, her dinnertime conversation with Michael concerned the Zimbabwean owner of South Africa's

---

[32] Informal bars, run out of residents' rooms.

Introduction

Mail & Guardian newspaper. More generally, my obvious wealth (evidenced by my pick-up truck) might have created distance between my informants and me. However, the workforce's diversity in terms of class aspiration mitigated this.

I participated in diverse aspects of compound social life, attended an apostolic church, became a fan of the workforce football teams, observed live music acts, and celebrated birthdays and New Year. Time after work and on weekends was crucial, when people caught up on news, exchanged gossip, and visited shebeens. During the day I followed work teams.

Making relationships with seasonal workers presented distinctive challenges, because of their much higher rate of turnover. I needed to get to know people quickly in a setting where many were undocumented and consequently wary. Becoming a member of a picking team enabled me to develop bonds with my teammates, while my obvious inability at speedy fruit picking made me more laughable than suspect. As the unpaid thirty-first member of a 30-man team paid on a piece rate, my labour was a bonus contribution to the group. Evening conversations with workmates offered further opportunity to acquaint myself with new arrivals. The transience of the seasonal workforce, however, meant that structured interviews with around 160 seasonal employees, focusing on basic personal information, were necessary to gain a wider sense of trends in the population.[33] This was later followed by comparable interviews with around 105 permanent workers and dependants, assisted by a male research assistant from the permanent workforce, from which I draw aggregated demographic data (referred to in Chapter 4). All of this was supported by small-sample surveys to address specific questions.[34]

As a male researcher, establishing relationships with compound women raised different issues. With women my own age (mid-20s), doing so provoked rumours of sexual interest. Nevertheless, there were exceptions. I was able, for example, to socialise with the compound's HIV/AIDS peer

---

[33] This was done with the help of five research assistants. Originally, I employed three men, with different roles at the farm. One was a permanent worker, one a seasonal picker, and one a volunteer of the HIV/AIDS NGO. The male skew was because of a difficulty in recruiting women, due both to a sense of female propriety among many, and to a lack of sufficient educational levels among others. I later recruited two women – Granny (see below), and a peer-educator volunteer from the HIV/AIDS NGO working in the compound. But most of the data was collected by four men in our twenties and thirties. This doubtlessly skewed the data, and its quality was further compromised by the number of research assistants. But this was unavoidable, as we gathered information on a large, highly transient population. Interviews with permanent workers used a revised questionnaire, employing only one research assistant, a male permanent worker. As one would expect, the data was far more consistent.

[34] On residence patterns in the compound; how seasonal workers had heard about employment and the date of recruitment; what proportion of people went home or remitted earnings after payday.

educators, whose reputation as occasional, overt sex workers put them beyond the realms of propriety. My participation in Michael's household enabled me to spend time with its female members: his partner, and his female relatives who stayed with him when they worked the harvest. I also worked with an older female informant whom I addressed as 'Granny', thus underlining a platonic and hierarchical relationship. Accompanied by her and thus fulfilling the demands of propriety, I was able to make visits to women and interview them.

My fieldwork was equally shaped by the sharp division along racial lines that continues to characterise many South African workplaces, particularly agricultural ones. Aligning myself more closely with one group automatically implied distancing myself from the other. Throwing my lot in with workers signalled that my views about racial/class distinctions were utterly different to those of farmers, who tended to look askance at me for my choice of residence. They saw compounds as dirty: as an inappropriate dwelling-place for a white person. The farm owner was probably too media-savvy to invoke notions of racial propriety in trying to dissuade me, but other white people in the area did express disbelief, and even disgust, when they heard where I was living.

This meant that, while residence in the compound ensured my immersion in its life, it set limits on the same kind of casual, everyday engagement with white farmers. They welcomed me into their offices or houses for tea or a meal, but tended not to invite me to parties, to church, or to farmers' meetings. Nevertheless, farmers continued to speak to me often and at length. Paternalist perspectives on race would even occasionally lead to a sympathetic view of my project. An older farmer, while he would never himself have lived in the compound, agreed with my approach to fieldwork. For him, immersion was the only way to understand 'African culture', of which he had a somewhat romanticised view.

Observing worktime interaction between farmers and workers was important for understanding farmers' roles and modes of self-presentation. I supplemented this with regular visits to farmers' offices or houses, where I conducted long, open-ended interviews, some of which I recorded. I also occasionally attended farmers' social gatherings. Farmer control over my access to their private lives, however, denied me any insight into the world of white farmer women. When I asked Willem's wife, Marie, if I could speak with her, I was told 'my husband can tell you everything'. White women's views, and their activities, remained a closed book.

Conducting fieldwork at a site of migrant labour is difficult, because of the linguistic complexity of a working population hailing from different places. I had begun learning ChiShona, the majority language of Zimbabwe, before I arrived. But as for TshiVenda, the predominant language straddling the border, I had to develop a level of competence

Introduction 33

during fieldwork itself. While an increasing number of Shona Zimbabweans were crossing into South Africa to make ends meet, many people from close to the border saw Zimbabwe's crisis as a problem caused by Shona themselves (the government was, and is, Shona-dominated). It was made clear to me, not long after I began residing in the compound, that addressing people in ChiShona as a default language was not appropriate. Making matters yet more complicated were significant minorities of SiNdebele, SeSotho, and XiShangani speakers, from parts of Matabeleland South further away from this stretch of the border. But these minorities would often employ the languages of others on the farm, in any case. Like many farm dwellers on the border, I attempted to patch together TshiVenda, ChiShona, and English as best I could, depending on the context.

A second difficulty when researching migrant labour is a lack of deep, everyday understanding of the sending context. I mitigated this by visiting Zimbabwe twice during the period of fieldwork: once with Michael, and once with Benjamin, accompanying them home and providing transport. Made during their paid leave from farm work, these trips were brief (a week in duration), and my hosts' circumstances as well-established, permanent employees gave a somewhat skewed picture of connections between Zimbabwe and my fieldsite. But they offered some insight into residents' circumstances at home, and we visited the homes of people who would later become seasonal workers. I had hoped to make further trips across the border towards the end of fieldwork, but was advised against it as tensions rose around the 2008 Zimbabwe elections.

Meanwhile, many workers move on towards Johannesburg and other South African cities after stints on the border farms. I addressed this by spending some time with a family that had left the farm to settle in Johannesburg, as well as visiting the Central Methodist Church, a refuge for many destitute Zimbabweans in the city (see footnote 3). Both my trips to Zimbabwe and those to Johannesburg broadened my understanding of farm workers' experiences. I do not draw extensively on them in this book, since the data lacks the robustness of my observations on the border farms. Instead, following the long tradition of labour studies in Southern Africa discussed earlier in this introduction, I make sense of Grootplaas residents' experiences in situational terms, in terms of life at the farm itself. Although not drawn on explicitly, the insights gained through my visits to Zimbabwe and Johannesburg inform my analysis.

I did, nevertheless, investigate the wider context of border farming. Like other compound residents, I moved between estates on the weekends to pay visits, watch football matches, and seek out music gigs. I also conducted extensive archival and oral historical research. Interviews, exploring the past and the present, were with farmers across the wider border region; with senior farm workers, and those with long personal histories in the border area; and with key local figures such as the mayor of 1980s Musina (the border town, then known as Messina), former Musina copper miners now residing in South Africa and Zimbabwe, and an amateur historian. All of these people's insights, directly or indirectly, inform my writing.

Since my fieldwork, times have moved on. As discussed above, Zimbabwe saw the establishment of a Government of National Unity, and an end to hyperinflation. In 2013, Robert Mugabe was re-elected as President in another questionable election. On the border, deportations ceased, and then started again. I have returned to Grootplaas twice since 2008. There, as anywhere, individual fortunes have continued to fluctuate, and people have continued to respond by making important life decisions. Yet much remains the same, with *mapermanent* central in the farm's mediated paternalism amidst the transience beyond the estates. Throughout the book, I note shifts in the wider context, and I return to the period since my fieldwork in the Conclusion. But, except when discussing events clearly specific to the period of fieldwork, I write in the present tense. I do so to capture the immediacy of what I describe, and to underline the fact that, despite constant flux, what is described in these pages is not consigned to the past.

The Structure of This Book

Chapter 2 introduces the white farmers of South Africa's northern border. It describes how they came to farm in the border area, and investigates what the farms mean to them. White farmers are often depicted as enacting a racialised civilising mission, but they are best understood through their own ideas about success and failure – in other words, what kinds of farmers they are. They emphasise individual personal effort, and see it as evidenced by the scale of their capitalist enterprises and their transformation of the landscape. Failure, an ever-present risk, is equally individualised, and equally personal. Today, keen to remain flexible in a climate of uncertainty and anti-farmer hostility, they strategise according to ever-shorter timeframes. But far from being simply practitioners of flexible capitalism, farmers have to lay down roots – literally – to succeed. While the future becomes increasingly uncertain – for farmers, and especially for their workers – in the short term, at least, the farms are based on provisionally stable, settled arrangements.

Chapter 3 constructs a history of South Africa's northern border area from the early twentieth century, drawing on archival data and unpublished, private photographic collections. Less an exhaustive account than a series of spotlights on key moments, it investigates the continuities in two ways of seeing the border: as a boundary to be enforced or as a labour-intensive hub of employment. It examines the attempts of emergent white capitalists in the border area – mine managers and farmers – to gain state recognition for the area as an important place in its own right, a centre of production, rather than a frontier zone. At stake was how the area was to be policed, and how employers on the border, often working at cross-purposes with police, would recruit southward cross-border migrants. The chapter, tracing these two ways of seeing the area to the present day, reveals centres of production as focal points on the border that organise and structure life for their residents. What emerges, therefore, is a long-standing tension between the needs of capitalist enterprises and those of the state.

Chapter 4 turns to the border's black residents, beginning with an exploration of life in Grootplaas's workers' compound. It reveals the centrality of permanent, male workers in diverse aspects of farm life. These established workers root themselves not only in contrast to, but also by means of, the transient border population. They build domestically congenial lives through their relationships with mobile women. And they enforce their version of justice through their associations with the soldiers who garrison the border.

The chapter examines how permanently employed workers assert their 'provisional permanence' by adapting their accommodation and establishing domestic arrangements. In a manner that echoes the experiences of black township and mine dwellers in Zimbabwe, South Africa, and elsewhere in the region, residents confront their insecure tenure by stress their rootedness. The extent to which they are able to do so distinguishes the permanently employed, who are overwhelmingly men, from the seasonal workforce, who are both men and women. Permanence is highly gendered: men use their virtual monopoly over access to housing to draw women temporarily into their own domestic projects. Labour hierarchy here is complex and multi-faceted, experienced through legal documentation, length of residence at the farm, and access to women for domestic labour.

Building on this, the following two chapters directly address how the effects of Zimbabwean displacement are refracted through labour hierarchies, at work and beyond.

Chapter 5 focuses on male seasonally employed pickers, showing how the work process – a fast, aggressive affair – is key in establishing camaraderie between new recruits. Belying the typical image of a short-term labour supply, however, many pickers have middle-class backgrounds or aspirations, having been driven to seek this form of work because of Zimbabwe's economic troubles. As they reject the workplace banter of their team-mates, they recreate a middle-class ethic that historically was defined precisely in opposition to labour migrancy. The argument illuminates how class is reimagined and contested through contrasting models of masculinity in the work process, under conditions of large-scale Zimbabwean displacement. This divergence shapes how, and to what extent, seasonal workers are able to become rooted in farm life.

Chapter 6 considers the interface between white farmers and black managers. It does so through a close examination of the two most senior black employees, the personnel manager and the foreman – their self-understandings, their life histories, and their place in the labour force and work process. Echoes of a regional history of racialised agricultural paternalism, combined with a growing rhetoric of corporate managerialism, produce sharp tensions in the workforce. As farmers abrogate paternalist responsibilities in an era of uncertainty and change, they shift towards corporate-style management delimited by contract. Such changes play out among workers. For senior employees base their authority on one or other model of labour relations, with their opposed notions of ethical responsibility to subordinates. Meanwhile, the shift to 'management' reveals how changes in agriculture intersect with those associated with the Zimbabwean crisis. The tensions between the foreman and the personnel manager reflect not only work dynamics and changes in farming, but also the different aspirational and class backgrounds of workers that owe themselves in part to Zimbabwe's decline.

Chapter 7 steps back from the labour dynamics presented in Chapters 5 and 6. It employs the insights of earlier chapters to reveal how farm-based arrangements shape the border more widely. It does so by examining the position and role of farm work in an economy of trade and services on the farms and across the border. The chapter underlines a central theme of the book – how border farm workforces' internal arrangements both reflect and refract the wider dynamics of the border and of Zimbabwean displacement. It does so by investigating the material and temporal articulations between ways of 'making ends meet'. This reveals how waged farm workforces represent key centres of gravity in the economic life of the border area, since even cross-border smugglers rely on their connections to the farms, and the accommodation and markets which these farms provide, for their livelihoods. Permanent workers wield considerable power, as traders need their say-so to reside in the compounds. For many traders, material accumulation may be less important than simply making enough to live on the farms, in socially stable environments where the alternative is insecure and unknown.

The **Conclusion (Chapter 8)** returns to black farm residents' 'provisional permanence'. It explores compound dwellers' vociferous assertions of belonging, as farmers have begun considering selling their estates to a multinational coal-mining corporation. By doing so, farmers underline their own capacity for mobility. For workers, this threat to the border farming area reveals the longer-term precariousness that accompanies their everyday stability – a key theme of the book. As the conclusion reminds the reader, labour arrangements on the border's white farms refract experiences of upheaval and attempts to achieve provisional permanence in an era of 'informalisation'.

2 'It's in Our Blood, It's in Our Skin'
Success, Failure, and Self-Sufficiency in Border Farming

Introduction: Enter Koos

In 1982, Koos, the Afrikaner founder of Grootplaas Estates, crossed the Limpopo River into South Africa. Having initially moved to Zimbabwe in the 1960s, he had decided to leave, unwilling to live under a black government. This was the aftermath of the 1970s liberation war, in which Koos had contributed his marksmanship and aeroplane-piloting expertise in support of Ian Smith's white regime. For the new ZANU government, the flight of white farmers meant potential economic disaster, including loss of the country's vast commercial agricultural infrastructure. One response was to prohibit the removal of farming equipment beyond Zimbabwe's borders.

Koos, however, was intent on moving his entire operation. He had seen potential in the land across the border in South Africa, on the southern bank of the Limpopo. Lacking sufficient funds for a large tract, he bought two portions of land in partnership with Jan, a farmer from the same Zimbabwean district. Jan saw similar opportunity on the border, and would join him there in 1983. He and Koos had agreed to split their farm after 10 years, once it was established, to ensure that their children would later be able to become farmer-landowners themselves.

Koos rented part of one of the agricultural estates on the Limpopo's northern, Zimbabwean bank. He planted a bit of cotton – enough to make this appear to be a genuine farming venture. Then he drove his agricultural equipment straight across the riverbed of the Limpopo from the Zimbabwean side, putting the word out locally that he was merely running the two farms simultaneously, with the same machinery. It was only through senior government contacts in South Africa, according to Koos, that he avoided being prosecuted for large-scale, illegal import after he had crossed. He paid the necessary South African duty but, he told me with satisfaction, the Zimbabwean government could not touch him.

One of Koos's neighbours today remembers his own part in the crossing. Thinus is the veteran of the border's surviving farmers. In his seventh year on the Limpopo when Koos arrived, he was already settled and knowledgeable about the locale. Sitting in the port-a-cabin he now uses as an office, he recalled:

They crossed right here... just about 500 metres from where we are now.... It was a lot of work, I think it took them about a week – between a week and two weeks.... It's a whole farm's equipment that they [brought over] – pipes, pumps, machinery, harvesting equipment, tractors – everything came through the river. One year out of all the years we know, the river didn't stop flowing, so you had to put wire mesh with our old cotton bags that they cut to join them together and put them on the sand so when the tractors went through they could not get stuck. I helped him bring it across, helping him putting the mesh, and helping clearing the south wall [bank] of the Limpopo River to make a gate where he could come through, and just helping with the operation.

That Koos's border crossing was not only resourceful, but also planned like a covert military operation, is perhaps unsurprising. Before farming and fighting as a counter-insurgent in Rhodesia, he had worked as a diesel mechanic there and in South Africa; farmed and fought against the Mau Mau insurgency in Kenya in the 1950s; and been one of only a tiny handful of whites living on the Caprivi Strip, at the eastern tip of Namibia, in the 1940s, when he floated logs down the Zambezi for a timber company. In Koos's self-understanding, tropes of soldier, pioneer, jack-of-all-trades, and bush hunter (of animals and people) combine. This was captured explicitly when he named his new South African farming operation *Houmoed*, an Afrikaans term used in the army, that translates as 'hold on', in the sense of 'hold it together a bit longer'.

Koos is seen as an inspirational figure among local white farmers, to whom he is known affectionately as Oom (Uncle)[1] or Oom KK.[2] The story of his arrival is widely known, and when farmers mentioned it to me they assumed that I too would know all about it. The river crossing stands as a shared narrative among the white farmers, a foundational legend that establishes the kind of people they are, and the kind of deed to which many might aspire. This tale is an example of what Rutherford (2001) calls 'pioneer stories'. Rutherford argues that farmers' 'pioneer stories' discursively shape how they think about their relationships with the land and its black population. He shows how farmers imagine themselves through their stories as the frontline of civilisation. They modernise the black residents of their estates while, toughened by 'primitive' life in the bush, they avoid the dulling trappings of modern, urban life. Such stories provide a useful insight into the ways whites speak about and practise farming, as this chapter will explore.

---

[1] *Oom* means uncle, and is used as a term of intimate respect for all older men in an Afrikaans community.
[2] These are the letters of his first and middle names. KK is used by black workers as well as white residents, generally in the context 'Baas KK'.

However, the chapter argues that such pioneer stories, when read in the context of farmers' everyday activities on their farms, yield a more complex picture. Farmers talk of transforming the bush into productive farmland because it signals their success – 'making it' – through sheer effort. Indeed, having come from Zimbabwe and from other parts of South Africa, it is precisely this full-time commitment to their work that they share. This chapter considers what the farms mean, as projects, to their owners.

Pioneer Stories

On South Africa's northern border, it is black Zimbabwean farm workers who have been seen as incomers, even intruders. But white farmers along the Limpopo River are also recent arrivals. Pioneer stories tell us something about how farmers regard their own mobile histories. They place individuated white men at the centre of an evolutionary narrative:

The genre of history... situated the story-teller and the farm on an advancing border of civilisation, bringing modern order to the bush... while at the same time it nostalgically invoked a simpler time compared to the hustle and bustle of 'modern times'. (Rutherford 2001: 82–3)

This self-imagination, argues Rutherford, is central to white farmers seeing themselves as fundamentally similar to one another, sharing racialised and gendered roles:

Pioneer stories emphasised the 'individual'... about them and other white men. They were the ones who hunted, who supervised the workers, who innovated with scraps, who 'opened up' the farm and 'made do'. When they talked about neighbouring white farmers, about the 'community' that developed between them in the face of common adversaries... they were always men. (Ibid.: 83)

I begin by showing the usefulness of pioneer stories for understanding how farmers think about themselves and their relationships. On South Africa's northern border, these narratives are central to farmers' self-presentation. Indeed, most are Afrikaners, unlike the majority of Zimbabwe's white farmers (such as Rutherford's informants). Their stories are inflected by an Afrikaner mythology of mobility: the image of the *voortrekker*, the proverbial Boer who migrated into the interior to avoid British rule (transposed today as the ever-mobile farmer, constantly moving to preserve his way of life); and more generally the *trekboer*, the pioneer farmer of the nineteenth century. Some border farmers, and especially Koos, draw on the trekking and/or Anglo-Boer War histories of their ancestors. These have become symbols of Afrikanerdom, centrepieces of the 1930s 'upsurge in the interest in Afrikaner history that would ultimately lead to the development of a distinctive Afrikaner nationalist school of South African history' (Gilliomee 2003: 432).

But 'pioneer stories' tell us a lot more about farmers than their models of racial difference or how they brought modernity to the 'primitives'. Among my informants, their stories were about what the *work* of farming means to them. In other words, to understand how white farmers think about their farming, rather than stop at focusing on the discourses of civilising missions and pioneering masculine isolation, we should also see farming in terms of the *projects* farmers strive daily to realise.

This opens up new questions. How do farmers *become* the men they aspire to be through their life projects? If farmers understand themselves as pioneering individualists, what does it mean to them to make a success of their enterprises, or to fail in this attempt? Rutherford moves the debate about 'white culture' along, by not assuming it to be a reified entity that causes beliefs and behaviours. He writes: 'I prefer to reverse this assumption and look at how these various discourses sustain, among other things, a notion of a unified "European culture" or "race"' (2001: 62–3). He relates 'pioneer stories' to the racialised spatial arrangement of farms, the focus of Chapter 4. His discursive starting point is also applicable to Koos and his farm, and thus frames the first part of this chapter. Border farmers' stories do indeed set them apart in terms of 'culture'/'race', often explicitly. But Rutherford's concentration on racial difference and modernising narrative may preclude an understanding of how farmers define themselves through the *successes* and *failures* of their enterprises, issues more easily explored by considering farms as projects.

Rutherford's racial focus is not unique, but follows wider trends in studies of white farming in the region (Zimbabwe, South Africa, and Namibia). Much agrarian research has emphasised the political economy of racialised paternalism (see, for example, van Onselen 1992; du Toit 1993; Sylvain 2001). This framework is indeed central to understanding white farms, and the relationships among their white and black residents. Farms on both sides of the Limpopo are racially divided places, total institutions whose totalitarian character has been reinforced historically by the exemption of farm workers from most labour legislation. Farmers' self-perceptions are equally racialised. Many cast themselves in the idiom of fatherly protector (van Onselen 1992; du Toit 1993; Rutherford 2001), while black workers are seen as children in an extended family, with senior male employees acting as intermediary 'paternal' figures (Waldman 1996: 69).[3] Much of this book will engage explicitly with these debates. But it

---

[3] In South Africa, such arrangements have been traced to the Cape's history of slavery, which formally ended in the 1830s but left an imprint on later Master and Servant Acts and the creation of a large landless rural proletariat (Waldman 1996; Ewert and Hamman 1999). In Zimbabwe, paternalism has a different genealogy. Nevertheless,

is also important to appreciate white farmers' own perspectives on their endeavours in broader terms.

New literature about Zimbabwean farmers places greater emphasis on farmers' imaginations beyond the question of racialised labour relations. This points to a changing stance towards white Zimbabwean agriculture in academic writing since its almost total collapse: one that demonstrates greater empathy than analyses of farmers in South Africa. Hughes (2010) seeks to understand how white farmers built dams, both to satisfy particular aesthetic sensibilities (approximating northern European landscapes) and to reinforce claims on territory by transforming it. Historically, both impulses have been means to assert belonging, and both do so *without* having to engage meaningfully with the majority black population. Before Zimbabwean land reform (beginning in the late 1990s), argues Hughes, farmers simply *ignored* the black population by focusing on the landscape.

But Hughes's interest remains the imagination of racial difference and, indeed, racism. This chapter explores farmers' points of view, concentrating on what it takes to survive in agriculture, and 'make it' on the land. A focus on life projects is similarly used by Hammar (2010) in her research on displaced white Zimbabwean farmers in Mozambique. She illuminates the enormous mental and physical effort involved in starting again, and the changing self-evaluations that accompany subsequent failure. Meanwhile, Pilossof (2011) shows how Zimbabwean farmers have long understood themselves in relation to the uncertainties of national politics. Attempts to ensure survival have led them to present themselves as 'apolitical' growers, focused on efficient production – something equally evident across the border at Grootplaas and the neighbouring farms. Going beyond a focus on race and the 'civilising mission', attention to farmers' projects means examining what is at stake in their ideals of success in life, and how these ideals connect them to the land.

Pioneer Texts

But let us first turn back to Koos's central place in local white imaginations, to elucidate what pioneer stories look like on the border. His life history conforms to the image of the archetypal pioneering white male in

---

Rutherford (2001, 2003) has explored a parallel, and similar, form of 'domestic government' on the farms, in which black working populations are considered the white farmer's domestic responsibility and are consequently excluded from any robust sense of national citizenship. Rutherford's Zimbabwean example questions the extent to which farmer paternalism, even within South Africa, really descended from Cape slavery. It appears, rather, to have more diffuse regional origins. These emerged from a range of close working relationships, including forms of tenancy and clientage, which only slowly gave way to wage labour (for Southern Rhodesia, see Phimister 1988: 84; for South Africa, see van Onselen 1996; Beinart 2001: 14). And they emerged from strategies by which small numbers of isolated white farmers controlled large, black workforces.

Africa: individually self-sufficient (when not actually alone); resourceful and multi-skilled; ever-ready in the face of adversity (natural and man-made); and handy with a gun. The romantic vision of Koos as hardy pioneer comes not only from his legendary crossing, but also from a book he wrote.

Koos's tells us something of how he presents himself. He had it professionally printed and bound in a glossy hard cover, and distributed it to his friends. Hunting is interwoven with stories of pioneering exploits, such as floating logs down the Zambezi and engaging in counter-insurgency operations, and the book ends with musings about evolution and God. White men are the main protagonists. The chapter in which black people come into the sharpest focus – *Mite en Legende* (Myths and Legends) – often refers to whole populations: the Basotho, or the 'Manika' (in Zimbabwe, Manyika, in Mozambique, Manica), for example. Black people generally feature as the manpower for white initiatives, and as the source of tradition, including folkloric yarns and an episode with a *sangoma* (healer). Photographs feature countless slain elephants and other big game, many with Koos and other khaki-clad men posing alongside, or feature Koos's wife and children, duly armed with rifles. There are also other images of Koos's exploits. In one series, he braves the Zambezi River in a canoe. In another, a white friend poses next to a timber truck. Many of the photos contain black men carrying out manual labour, but never as the primary object of interest. One, for example, depicts around eight men using poles to punt a raft across the Zambezi; the real focus of the image, however, is the car on board, suggesting civilisation in the bush.

Koos's tracking expertise in the bush and hard militarism, part of his pioneer persona, are evident in the book's design. The cover features an elephant he shot as a young man, rendered in the style of a watercolour painting in sepia tones. The back cover displays a photograph of the same elephant, with Koos and his wife sitting embracing on top of the corpse. He is wearing a khaki shirt, shorts, and socks, his wife a light dress and sun hat. The words 'Kilo Sierra' under the photograph are the radio code words for the initials of his name, evidencing a similar military aesthetic to that suggested by the farm name Houmoed.

Koos's book is one of several written sources that white residents of the border read with interest as windows into the local past, and that they use to underline belonging. The content does not portray the area itself. Instead it paints a picture of a key resident, and strives, according to its author, 'to give the younger generation a look back at a bygone era in which their elders lived' (*'an die jonger geslag 'n terugblik te gee op 'n vervlöe era waarin hulle voorouers geleef... het'*; spelling in the original). Other historical sources are exchanged and circulated. They include the records of a parliamentary commission detailing land use in the area in the 1940s, and the memoirs of the wife of the English-speaking manager of the local Messina copper mine, who between the 1920s and 1970s owned the land that Grootplaas now occupies. Such books enrich whites' sense of belonging, as they gather detailed information about the area and act as guardians of precious historical sources. In a sense, they become guardians of a particular – white – version of local history. Indeed, one white amateur historian from a nearby town removed the early twentieth-century registers from a mission hospital – a national heritage site – because they contained his ancestors' names, and he felt 'his' history was not being conserved with the care it deserved.

Such written sources are incorporated into local pioneer stories. White residents are knowledgeable about who the oldest families are (cattle farmers dating from the early twentieth century), who has the best historical expertise, and who the best collections of documents and photographs (one such collector is an estate agent from nearby Musina, mayor of the town in the 1980s, who appropriated a huge collection of photographs and records from Messina Mine when it closed its offices). When I spoke to white residents – from the border to Louis Trichardt, south of the Soutpansberg Mountains – the same names kept coming up. For many local white residents, a degree of detailed knowledge about the past asserts their belonging in a place to which people came in scattered groups and in different periods, some only in the 1980s or since.

This self-consciously preserved settler history distinguishes the area from that described by Rutherford (2001). Of Zimbabwean farmers he notes that they 'would likely find my application of the term "history" to their narratives concerning the "past" inappropriate since they lack the wealth of empirical detail that, among other things, helps sustain a notion of "history" as an objective recounting of past events' (2001: 82). In the Limpopo borderlands of South Africa, white residents – farmers and others – take a keen interest in 'history', giving attention to apparently objective details. Their pioneer stories are peppered with such minutiae as exactly which farmers arrived and which went bankrupt in particular years, and in some cases which families were intermarried with which of the older white dynasties. I learned from another farming couple, for example, that Koos's son was married into one of the old English farming families whose estates lay to the southeast, as they chronicled the various marriages between lineages, including their own.

Assertions of historical knowledge are often present in farmers' stories. Koos's book is a source that his friends use to add detail to their historical consciousness. It is also a classic pioneer story in its own right, squarely in the narrative genre elucidated by Rutherford. Its wealth of historical detail is matched by its relentless focus on an individualised 'I', along with its themes of resourcefulness, bush-readiness, and understanding of black 'myth'. However, Koos's book was written as an account of his larger-than-life exploits, and is focused on the exciting highlights of the author's life. The stories told by farmers in person, by contrast, offer a more complex picture, in which the individual self-sufficiency of pioneering masculinity is applied to farmers' everyday concerns, namely their farms.

Men of Vision

In response to my questions, farmers would often describe the development of their farms. They would present this development as the result of their own individual effort, perhaps with the occasional help of another farmer. Recall Thinus's version, above, of Koos's river crossing. It portrays the moving of a whole estate's worth of equipment across a soft, sandy river bed (and national border) as the work of a few people: 'I helped him bring it across, helping him putting the mesh, and helping clearing the south wall [bank] of the Limpopo River to make a gate where he could come through'. Despite the scale of the operation, in its retelling there are only comradely white farmers in sight.

Thinus's description of his own farming history further exemplifies such self-imagination. He tells how he began planting cotton on the Limpopo's south bank in 1975. When he came to the area, he reports, there was little crop farming, with 'very poor irrigation schemes', and much of the land was undeveloped or used for cattle. Thinus bought land from an absentee landowner based in Harrismith (in KwaZulu-Natal) and 'opened it up', buying further portions of land as he expanded production. According to Thinus, land along the Limpopo is very fertile, and conditions are perfect for farming. But the area's remoteness, lack of basic infrastructure in the early years, and the fact of starting from nothing in the middle of a low-level border war drove out all but the most serious would-be farmers. Speaking to Thinus and other border farmers, the picture that emerges is a pantheon of individualised, white, male farmers making their own way, helping one another when necessary. This version of the area's development is reinforced by the black farm workers' practice of referring to farms by the name of the farmer, prefixed by the vernacular term meaning 'place of'. 'KhaDeki' is the 'place of Dirk', who went bankrupt and now manages Jan's estates; KhaRudi refers to one of Koos's sons, Rudi, who now runs their second farm near Komatipoort; KhaGideon and KhaKobus refer to the farms' current farmers, Thinus's two brothers.

Of course, narrating the area's development as the product of such pioneering enterprise does not automatically valorise hierarchies of racial superiority. Some farmers did however make explicit such a framework of interpretation, one with strong paternalist overtones. As Koos spoke to me during my teatime visits to his house, it became clear what being a farmer meant to him. Crucial to this was a fundamental difference between 'European' and 'African', the former characterised by far-sighted vision, the latter by a level of instinct that Europeans have lost by their exposure to 'civilisation'. This contributed to his feeling that his role as a farmer was that of a visionary, a pioneer, and an embodiment of modernity, but another frequently emphasised trait was his expertise on 'the people', in this context meaning black labour. He claimed that, for this last reason, Afrikaners were especially in demand by the World Bank for the reconstruction of agriculture in post-conflict Mozambique. In other words, farmers like Koos are the frontline of civilisation, but sufficiently in touch with the 'bush' and its people to understand how to command authority there.

Koos would demonstrate such understanding in his anecdotes and in everyday speech. He jokes with workers about Africans' need for another firm hand like that of Shaka, the nineteenth-century Zulu king; he commented to me that workers had appreciated his buying a new four-by-four car as, for them, leaders without the right status symbols reflect badly on their followers. Forged in the context of racial inequalities, work situations, and colonial and apartheid rule, Koos holds a particular, romanticised ideal of 'the African'. In his accounts, warrior metaphors, superior skill in tracking animals, and superstitions about witchcraft jostle with his assertions of the appreciation of such people for firm authority.

Other farmers in the area would speak in a similarly philosophising tone, even building theorised models of racial difference. On more than one occasion, Jan, Koos's neighbour and former farming partner, spoke to me about the difference between 'ice man' and 'tropical man'. The former, having evolved in the cold, had learnt to save food for the winter, keeping Chicken Licken meals (a joking reference to a South African fast food chain) in his cave. By doing so, he learned to plan. Tropical man, meanwhile, had never lacked access to food through the year, and so had no need to plan. The cultural models are highly gendered, evoking contrasting masculine hunter-gatherers. Though invoking different metaphors, Jan's point was very similar to Koos's: Europeans are distinguished from Africans by their 'vision'. Gesturing on one occasion to his Toyota Hilux pick-up, Jan told me that this was a product of ice man's cultural advantage. He added that tropical man also has his strong points, but it took him some time to think of any. 'Socialism,' he eventually concluded. Koos shared Jan's belief in Africans being more social. For Koos, the strong point of African culture is their diligence in looking after the very young and the old. The culture of the African, or 'tropical man', is the antithesis of individualistic modernity, with its advantages and disadvantages. Another farmer, John, philosophised for an afternoon on the difference between African and European religious outlooks, in terms of their effect on individual responsibility.

It is clear, then, that farmers' stories and musings present white, male farmers as the protagonists in rural development and progress. As Rutherford points out, such narratives reinforce a sense of white community. They do so all the more effectively when some of the stories are sensational and are retold again and again, such as the recounting of Koos's river crossing. Pioneer stories often concern the role of farmers' European 'vision' in developing their agricultural estates.

At the same time, farms illustrate the success of work-intensive projects in farmers' assessments of their own worth.

Pioneering Projects

Farmers' stories about bringing civilisation to the wilderness are equally claims about the personal meaning of work. Such claims can be found in other settler narratives, such as those from the United States. 'For American newcomers, settled farming held moral as well as physical meaning,' notes Valencius, in a history of nineteenth-century American settlers' conceptions of the landscape in relation to health and disease (2002: 195). She explains that 'agricultural labour was not merely labour. It was an engagement with the natural world in which cultivators imposed their own will, putting an end to "the primitive state of nature"' (ibid.: 198).

There are important historical differences between agricultural settlement in America and in Southern Africa. But in both, the wider resonances of being 'a planter of rows' (ibid.: 196) placed white settlers as frontiersmen of civilisation, rationalising an environment through their vision. To appreciate what this means for border farmers in northern South Africa, however, means moving beyond contrasting white 'modernity' with black 'tradition'. I turn now to investigate how farmers speak about and practise the work of transforming the landscape into a series of profit-generating farms. Pioneer stories are not merely elaborations of a civilising mission; they are also a way into appreciating the sheer effort farmers feel it takes to transform their environments.

The transformation these farmers have achieved as they built their enterprises is considerable. The border farms lie at intervals along the South African side of the border, their fields, orchards, and labour compounds stretching away from the road that runs along the fence (see Figure 2.1). Jan, with 1000 hectares under cotton and a further 300 under citrus, claims to be the largest cotton producer in South Africa. Thinus, with a little less than 1000 hectares in total, claims to be the largest supplier of tomatoes for a household brand of ketchup. Koos's Grootplaas reached an output of one million crates of citrus in 2007. Its fruit is sold in several major supermarkets in the UK, and across Europe and Asia. Its lands comprise around 450 hectares under orange

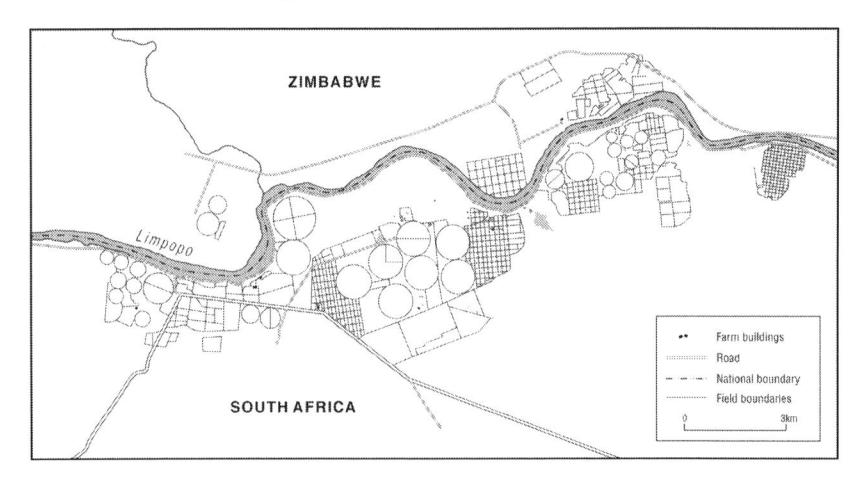

Figure 2.1. The border farming area during the period of fieldwork. Circles are pivot-irrigated fields for crops like cotton, rectangles citrus orchards or tomato fields.

and grapefruit orchards, and a similar hectarage of uncultivated *mopaneveld*.[4] Such farms require extensive infrastructure for irrigation: Grootplaas has a private, lake-sized dam (reservoir) for the citrus trees, and during fieldwork a second, larger dam was being constructed. These are filled during the two months that the Limpopo River flows, using pumps on the riverbank.

Farming here is, indeed, hard work. It is puzzling how much effort a white farmer-landowner needs to expend when one considers that he has a workforce of as many as 600 workers. Being managers, however, little of what farmers do involves manual labour. In this respect white farms are little different from many other settings of large-scale, organised production where class rather than racial division prevails. On farms, however, everyday organisation makes white farmers' roles especially clear and visible. Established ideas about the division of labour, alongside the practicalities of work across hundreds of hectares, place farmers at the centre of their operations. By contrast, in American factories for example, management has to carve out its own space, imposing rules that legitimate its presence but that hinder efficient production (Burawoy 1979).

The work activities of Willem, Koos's son-in-law and farmer at Grootplaas, provide a good example. He is at the farm offices each morning by 6 a.m., and during the harvest may stay at the packshed overseeing work

---

[4] According to hectarage, Grootplaas appears smaller than its neighbours, but this is misleading. A hectare of citrus trees represents far greater investment and output than the same area of cotton.

until 10 p.m. His day involves negotiating with export agents; liaising with inspectors from South African state institutions, foreign supermarkets, and EUREPGAP, a European standards agency; and investigating and buying agricultural inputs. As in many other industrial enterprises, while various dimensions of production occur relatively independently of top management, the organisation as a whole is deeply hierarchical. In particular, there is a bottleneck in communication between people involved in production on the farm and outside organisations, whether state or private. Much of this communication is currently channelled through Willem. His work involves speaking to representatives of institutions with which Grootplaas has dealings, as noted above. As we shall see later in the book, farmers like Willem are changing how they speak about and enact their roles, as they attempt to fit a corporate organisational style. It is becoming advantageous for farmers to promote black front-men for their enterprises, to avoid allegations of racism. Caused by changes in the export-agriculture economy and in the politics of farming, this privileges some workers, notably clerks, and threatens others, such as foremen. At present, however, the major flow of communication with buyers and state officials runs through Willem.

There is another area of work that occupies farmers, that of the apex of a chain of command. Farms have peculiar kinds of hierarchies in which nodal points are of key importance. On the one hand, workforces are often dispersed over huge areas. On the other hand, the hierarchy extends beyond work issues, narrowly conceived, and is based on expectations of 'farmer paternalism'.

Black foremen and white managers are key to organisation in such circumstances, maintaining contact with their two-way radios, acting on their own initiative, and representing the farmer on the ground. Willem's wife, Marie, is a crucial go-between, from her office at the workshop. And the packshed during the picking season has its own pecking order. The offices where Willem and his family work represent a key hub in this kaleidoscope of operations.

This steep, centralised chain of command affects all aspects of life, because workers live on-site. They come to the farmers with a range of complaints, including marital disputes and tales of weekend violence. This dimension of the labour hierarchy is a significant theme in later chapters. For now, it is important to note that, for some farmers, arbitrating between workers with respect to their non-work lives is part of a paternalist vision of farming. In other words, this is the practical side of the tales of civilising mission we saw earlier. For others, like Willem, having to intervene in the lives of up to 600 workers and their dependants is the result of established expectations, and it is an annoyance. Willem's language for describing this work is 'labour relations', hardly the rhetoric of paternalist responsibility. And he reinforces his perspective by claiming not to understand enough about his workers' 'culture' to make sensible decisions about their private affairs. Here, therefore, farmers differ over their proper relationship to their workers; not all fit the model of the modernising pioneer with paternalist responsibilities. All, however, have to engage with established expectations. And this is a major part of their work.

At the same time, farmers are never all that far removed from the everyday tasks of the farm. I would sometimes see Willem not just directly supervising repair work in the workshop, but on occasion even getting under a tractor to fix it himself. In similar vein, Thinus proved to be extremely hands-on when it came to policing the well-being of his crops. I first met Thinus when I was visiting his white neighbour. It was evening – long after dark – and Thinus stormed up to the house to find out who had sprayed dust over his tomatoes by driving too fast (I was the guilty party). He had been out on his tractor at the time, still in the fields even after his workers had retired for the day. Lying behind white farmers' perceived need to keep a close eye on everything is their assumption that it is they, not black workers, who have the necessary vision to farm. The pioneer stories with which we began contain models of racial difference that shape farmers' experiences of their work. In any case, when farmers describe themselves as jacks-of-all-trades – simultaneously businessmen; soil, crop, and chemical experts; overseers of labour; mechanics – there is significant truth in what they say.

The farms are, after all, economic projects to which their owners feel deep personal attachment. Thinus expressed this especially clearly. To explain how he came to farm on the border, he felt he should start by explaining: 'We Westhuizens are farmers. It's in our blood, it's in our skin. So, for that purpose, I also went into farming. My brothers also all went into farming.' He went on to describe why he moved here, in 1975, to plant his first crop in 1976:

> I visited the area here, one time in my life, and I saw the opportunity of the best climate, soil – alluvial soils that's been put down by floods from the Limpopo River – and good quality of water. So the very first chance I had – I worked for a boss – I came here, hired a small farm, then I got an opportunity to open up the bush, take the trees out, and started opening up this farm. And from there on when I developed the whole farm I bought the neighbours' smaller farms, five of them, until I own six farms today.

Although this statement provides ample evidence of the masculine individualism of the pioneer story, it also shows something about the importance of sheer hard work. The lone white man 'opens up' the bush, which was, in Thinus's words, 'like Africa'. But the transformation achieved is the result of hard, committed effort, and this is what Thinus is keenest to underline. For a farmer like Thinus, *how* he works is central to his self-understanding. 'I am one of the difficult people in life,' he reflected to me. 'I'm a perfectionist. I will do a thing right or I will not do it. If I open a farm I will not open a bad farm. If I've got a choice to open a farm in the best area I will not open it in a worse area.' The 'best area' is a place of good long-term potential – good climate, soil, and water – not the easiest or the most comfortable location. Thinus highlighted the border's challenges. Infrastructure – electricity, telephones, upgrading the rough dirt roads – only arrived in 1984. Fewer than 20 per cent of the original white farmers from the 1980s survived. Most went bankrupt or lacked the necessary commitment to their farms to persevere. The really serious farmers, those who made it, did so because of their hard work. This Thinus emphasised to me, stopping the interview to ensure it was included in the recording. Speaking with slow, measured words, he articulated why so many failed. His thoughts are worth quoting at length, as they state clearly what I heard in different forms from farmers along the border:

Although the opportunities here if you know what you want to farm and how you must do it is [sic] the best in South Africa – the yields we take from crops here is higher than any other place in South Africa – but the difficulty to do it is much more than other places because you've got to take, our water we take out of the sand of the Limpopo River, not from an open stream, and it's a very hard area. December, January, February is so hot that some people can't take the heat. So you must know how to make use of that, to benefit out of it. Farmers that messed up, that are not here today, thought it's very easy. One farmer came here; it looks like he's doing well. The government is helping to get a loan to come here [in the 1980s], it'll be very easy. They didn't know that this area is a difficult area.

Thinus is an extreme case. He emphasises his hard work, even above other farmers in the area. He underlines the seriousness of his operation by reporting that even his wife drives tractors and is involved in the planting – something unusual in farmer self-representations and practice. Whatever his pitch for distinctiveness, the basic self-understanding that Thinus conveys is more widely shared. The idea of the pioneer not only foregrounds individualistic white men, it also emphasises work. And border farmers constantly contrast themselves with those who did not make it. Here is Jim, a much smaller farmer who owns one of Thinus's land portions in partnership with him:

If you go back into the backgrounds of people here, what were they? Schoolteachers, soldiers, and all the other professions that you can think of. It was actually a minority of people from farming backgrounds that developed this area. So it was 'open the land, clear the bush, now we've got to establish water'. There's not a tap that you open [and] the water runs out. You've got to go and find that water. . . . This is all pioneering work that was done by the farmers here. So it was survival of the fittest. And the weaker ones have gone, there's no two ways about that. People didn't all make the grade here. Hence the smaller guys have always been bought out by the bigger guys as it's become more developed.

It is striking that even Jim, without doubt the smallest landowner in the area, is so keen to romanticise the success of the hard-working pioneer. What the largest and smallest farmers along the border share is a sense of personal character rooted in the valorisation of effort. There is a sense here of what Katherine Newman, in her study of American middle-class downward mobility, calls 'meritocratic individualism': 'At the centre of this doctrine is the notion that individuals are responsible for their own destinies. This idea, which owes its origin to Calvinist theology, carries into the world of work a heavy moralism' (1988: 76). Perhaps unsurprisingly given their commitment to Calvinist churches such as the Nederduits Gereformeerde Kerk (Dutch Reformed Church) and the more conservative Afrikaanse Protestante Kerk (Afrikaans Protestant Church), a similar moralism or 'protestant ethic' (Weber 1992) weighs on border farmers' self-understandings. As a farmer, success and failure in agriculture reveal who you are.

'I'm Here to Farm': The Serious Business of Border Agriculture

A serious, hard-working approach to agriculture is not only a means by which border farmers evaluate themselves and one another. It offers a way for the border community as a whole to compare itself to others. Cattle and game farmers 50 km behind the border have established an armed security force called Farm Watch as a measure against violent attacks on white farmers. This organisation was behind a number of news reports in 2007 featuring farmer vigilantism on the border. After I attended one of their self-defence camps, which involved learning to use a range of firearms, I asked Paul, one of Koos's sons,[5] his opinion of these camps. He referred to the organisers as 'weekend warriors', people with too much time on their hands because they have too little real work to do. For Paul, as for other border farmers, what distinguishes them from the cattle and game farmers – who preceded them in the region, and whose land surrounds them to the south – is that border farming is the real thing. The scale of the enterprises and the work involved define them.

Border farmers have various explanations for their exceptionalism. For Jim, this is a pioneering story, whose protagonists were young at the time, and 'more . . . aggressive than the guy who's going onto a developed

---

[5] During fieldwork, he was responsible for Grootplaas's water infrastructure. He has now bought a nearby guesthouse and left farming.

piece of land, [where] it's all peaceful and there's settlements'. An apt case for comparison, in Jim's eyes, is that of America's cowboys. Youth and aggression explain the border farmers' drive to work, and distinguish the successes who remain from the many failures who left. Jim's version of border success resonates clearly with the 'pioneer stories' discussed earlier. Here are romanticised, highly masculine figures, beating back the bush: the individualistic pioneer ideal.

Other farmers – especially those with backgrounds in cultivation – have quite different interpretations of the area's success. For Jan, what made the border such a serious agricultural area was the large proportion of *Zimbabwean* farmers:

We [Zimbabweans] weren't a politician or a businessman or what-have-you, or a teacher that owned a farm.... The whole country revolved around farming. So the best of the best were bred there, amongst this farming community. Business-wise and farming-wise and how to make success out of farming. That was the culture of the country. Now, if you look at Zimbabweans who moved out of Zimbabwe, you'll find them – a lot of them, around the world – very, very successful. Very successful. The biggest tobacco farmer in South Africa is a Zimbabwean. The biggest fruit/vegetable [farmer] is a Zimbabwean. The biggest cotton farmer [Jan himself] is a Zimbabwean. And then you look at the citrus....'

For Jan, the point is *not* to become too enamoured of the wider meanings associated with being a settler. We saw earlier that he has detailed and carefully articulated theories of racial difference (ice man versus tropical man). His theory explains, for him, why white people should be running farms, and represents the kind of thinking that lies behind pioneer stories. But being serious about farming, for Jan, means focusing on the job at hand, not conflating different activities and masculine ideals. Unlike Koos, for whom being a farmer, a soldier, a hunter, and a mechanic were integrated into being a pioneer, Jan was clear that settling the border was, in his case, about finding good land and planting it. His objective in Zimbabwe was 'to farm, not to fight the war', and this attitude towards his enterprise continued in South Africa: 'I'm here to farm, not to get involved into politics or the war or anything.' He believes that South African farming stagnated historically, becoming tied up in complicated ideas about being a Boer. Consequently, South African farmers lacked skills that Zimbabweans had. The border, says Jan, has benefited from the influx of professional-minded, focused farmers from the north.

Matters are far more complicated than a story of Zimbabwean farming, displaced to South Africa. Jan's statements themselves have a particular motivation. Farmers like Jan hope to avoid 'politics' in post-apartheid South Africa – unlike their more conservative counterparts in Farm Watch – by presenting themselves primarily in 'economic' terms, as businessmen. This resembles the 'affirmative parochialism' that Pilossof detects in farming history in Zimbabwe itself (2011: 70). The view among the border area's Zimbabwean Afrikaners is that they grew up more broad-minded than their southern counterparts because they learned both Afrikaans and English 'culture'. That is why far-right politics does not flourish along the Limpopo, unlike on the cattle and game farms away from the river – and why, by implication, the border farmers should not be considered a problem by the South African state. Chapter 6 explores this 'business' style among border farmers in detail, and its consequences for workers. For now, it is enough to note that a 'modern' self-presentation – being a serious, business-motivated capitalist, not a traditionalist Boer – has perceived political benefits in the new, somewhat anti-Afrikaner South Africa.

The reality, of course, is less clear-cut than this opposition between professional farmer and pioneer suggests. Neither do such differences neatly reflect an opposition between Zimbabwean and South African farmers. To start with, Zimbabwean farmers' outsider status is belied by South African genealogies, even upbringings, and they acquired South African citizenship when they began farming in South Africa. Unlike their workers, to whom I turn in the following chapters, their mobility was underpinned by secure, rooted legal status. Their Afrikaner backgrounds further mitigated any sense of difference from like-minded 'serious' farmers in South Africa, such as those at nearby Tshipise.[6] Jim and Koos, both of whose experiences of farming were shaped in Zimbabwe, also see themselves in the pioneer mould. Thinus, a South African, subscribes to the pioneering ideal, but is also the most 'serious' agrarian capitalist in the area. Despite the varied texture of their self-understandings, what is clear is that the border farmers have together developed a competitive, effort-focused sense of farming as a transformative life project, whatever its other associations.

Whether as a result of their pioneer aggression or their Zimbabwean approach to agriculture, border farmers see themselves as differing from other, less-intensive farmers. What stands out in both these self-assessments is their capacity to work – to build their lives around commitment to their enterprises. In both, farmers see themselves as lone operators – self-sufficient individuals building life-defining projects. But ideals of self-sufficiency and individualised success are a heavy burden, as the next section reveals.

The Enterprise Strikes Back: The Burdens of Self-Sufficiency and Success

Thus far I have considered 'pioneer stories' and their role in farmers' broader self-understandings. I then took matters further, by enquiring how farmers think about work in their agricultural projects. Appreciating

---

[6] See next chapter.

this latter perspective enables us to explore how farmers experience a sense of individual effort and self-sufficiency not merely as the stuff of romanticising narratives, but as framing interpretations of success and failure. These have real, everyday consequences, as this section shows. Sons struggle to fill the shoes of their fathers. Meanwhile, the accomplishments of the border farmers are underpinned by white managers, whose roles are a constant reminder that they have failed to acquire farms of their own.

First, however, it is necessary to understand how farmers' individualised self-understandings are rooted in a way of living that extends beyond their work, and to which the new generation must commit themselves. When Koos explained to me what it takes to be a farmer, he emphasised being a 'loner' – being comfortable in one's own company. Sometimes you do not see another person for days, he claimed. That is why he sent all three of his sons away as young men, before he would let them farm. Their choices to farm had to be informed by knowledge of the alternatives. Paul, for example, studied mechanical engineering. The point was that they knew what life was like in society, yet *still* wanted to live in rural isolation. You cannot have a farmer running for the 'bright lights' at four o'clock every Friday afternoon, Koos said. He recalled with pride how another son, Rudi, who now farms the family estate at Komatipoort, came home whenever he had the chance, to 'sit on the tractor'.

As with many of Koos's reflections, this sounds like an exercise in self-imagination as a prototypical pioneer, with life lived in the absence of other human company. Certainly there is a degree of exaggeration and romanticisation in these comments. They also betray a sharply racialised sense of proper sociality. Crop farms have large black workforces, and white residents all have domestic workers and gardeners. Not seeing another person for days refers to a lack of *white* company. Koos's views reflect both a strong sense of racial difference and a taste for hyperbole. But they also point to the fact that farmers' sense of individualism must be understood in relation to their experiences of everyday sociality.

Border farmers live in farmhouses on their land, each with his nuclear family, at a distance from the compounds in which black workforces reside. Each house conforms to an aesthetic of well-tended, lush gardens, replete with flowers, trees, and lawns – all surrounded by a security fence. Koos's is particularly striking because part of the house is built around a large baobab tree, which therefore appears to emerge through the roof. Interiors vary markedly: Koos's living-room furniture is wooden and expresses a *voortrekker* theme, with the arms of chairs and a sofa fashioned as wagon wheels; Willem, his son-in-law, has a living room dominated by a large fishtank and a flat-screen television.

Fathers, sons, brothers, and white managers live in similar farmhouses scattered across the estates. Even such houses on a single farm are distributed far apart. White residents travel around in four-by-four cars or pick-up trucks. The fact that many houses must be accessed via the dirt tracks that criss-cross estates means that chance encounters among farmers from different estates are relatively rare. All border farmers use the road along the border fence as an arterial route. But because, on a daily basis, farmers only move around their own farm, each drives up and down a different segment of the same road.

Given this distribution across the landscape, the fact that farmers spend most of their waking hours working means that they do not see each other very often. When I was conducting interviews, I was struck by the ways that border farmers defined themselves, as a group, in terms of their shared experiences of arrival, establishment on the land through their work, and seriousness about farming. I asked about community, looking for ways that farmers considered themselves bound together socially. But such inquiry was repeatedly met with explanations that they have no time to socialise. As we saw earlier, it is farmers' assumptions about their roles on agricultural estates – assumptions evidenced by pioneer stories and theories about racial difference – that lead them to feel they have so much work to do. And, as Thinus pointed out, success itself created the scale of work involved.

> After it [the land] was bought out it belonged in larger areas to less farmers, like a snowball effect. The more land you work the less time you've got to run around, so today everybody's doing his own thing and you can be so busy that you don't see your neighbour for six months.

There used to be a large number of families; each farmer initially bought only one or two portions of land along the river. And the farmers were young men, with young children. Half a dozen families now largely dominate this stretch of the border. Thinus, for example, owns six portions of land, and Koos and his sons four. The expansion of farmers like Thinus, Koos, and Jan, and the bankruptcy of others, has meant over time that there are fewer white people around with whom to socialise.

White residents of the border estates remember parties in the 1980s, for Christmas and to celebrate successful cotton harvests. One farmer reminisced about fireworks displays in the Limpopo riverbed on Guy Fawkes Day. Doubtless, these accounts of the past are inflected by nostalgia for a lost era, when farmers who are now middle-aged were young men, and the up-and-coming generation were children. It is certainly true that border farmers continue to organise events. Members of Koos's and Jan's families play polocrosse.[7] They hold tournaments, for which teams travel considerable distances. These, and the practice games that lead up

---

[7] A hybrid combining polo and lacrosse, it is known locally as 'poor man's polo' as it only requires one horse per player per match, rather than one each chukka (game period) as in polo proper. It involves throwing and catching a ball using lacrosse-style sticks, from horseback, and scoring goals between posts at each end of a rectangular playing field.

'It's in Our Blood, It's in Our Skin'

to them, are important opportunities for socialising. Tournaments last whole weekends, with a *braai* (barbecue) and drinks at the *lapa* (barbecue enclosure) on Saturday night. It is often the same families that attend, so durable networks are maintained through the games. These, however, are not frequent occurrences. They take a lot of effort to organise, and divert energy from the work of farming itself, which normally includes Saturdays. As far as everyday sociality goes, there has thus been a marked change.

Church services provide one opportunity to see people. The *dominee* (minister) of the Musina Dutch Reformed Church conducts a service in one of the border farmhouses once a month, rotating between the families. Other weeks, farmers and their families may drive into town for the Sunday service at the church itself. But the split between the Dutch Reformed Church and the Afrikaans Protestant Church (which occurred when the former allowed black members of congregations and homosexual *dominees*[8]) has complicated this opportunity to meet. Meetings of the local farmers' association may be accompanied by a *braai*. But, taken together, these various get-togethers are a very limited source of sociality.[9]

This takes a particular toll on the young. The son of one farmer, now in his twenties, spoke to me of how there are very few people with whom he can spend time. His twin brother, uninterested in farming, has left. In 2008, he himself had recently completed a course at an agricultural college near Nelspruit, not far from Swaziland. There, he was surrounded by people his own age. When he returns to the border after seeing them, he says, it takes weeks to become reaccustomed to the isolation. He has a beer sometimes with his cousin, Klein Thinus (Little Thinus – Thinus's son). But Klein Thinus, like many of the 'younger generation', is busy taking over running estates, and these younger men are starting families. For those who themselves farm, there is little time for anything else. For those who are not so busy, the border can be a lonely place.

Koos's grandson – Willem's son – is at present confronting this dilemma. Jacques wants to be a farmer, but is becoming aware of the

---

[8] Following the end of apartheid, and as a result of pressure from the ANC government. Of the Afrikaans churches, the Dutch Reformed Church is by far the largest; smaller churches, less symbolically important, were not induced to change after 1994. The Afrikaans Protestant Church was a breakaway from the Dutch Reformed Church by members who opposed the new reforms, and who feel that the Church was erroneously seen as whites-only; the important distinction, they say, is 'cultural' rather than 'racial' – whether or not someone is an Afrikaner (and therefore white). On the border, some farmers, including Koos and Willem, shifted allegiance to the Afrikaans Protestant Church in response to the reforms – even though the local DRC *dominee* is known to share their views, and there is no chance of either mixed services or homosexual ministers in the area.
[9] I was not able to attend services, because of my limited access to farmers' private lives, as described in the introduction.

realities of the isolation involved. When I began fieldwork in 2006, he was in his mid-twenties and finishing his studies in agricultural economics at Stellenbosch. He returned from university with styled hair and board shorts, cast in the image of a surfer. Keen to ensure his son had options in an uncertain future, Willem sent him to Belgium for nine months, where he worked for one of the agents that handle Grootplaas's exports. When Jacques returned, he had made up his mind. He had found Europe cold and alienating. His intentions were reflected in an abrupt change of style: buzz cut, khaki shorts and shirt, walking boots, and an underarm holster for his handgun. Willem bought him a pick-up truck, and organised for him to manage a farm up the road that had been bought by one of the export agents. The arrangement was ideal, Willem explained. Jacques would run the farm himself, but Willem was close by whenever he was needed, to offer guidance. Jacques could also benefit from Grootplaas's existing procedures, like those for picking up the large sums of money required for monthly paydays.

Jacques was very excited by the opportunity. When I visited him at his new home, he showed me around his packshed, recently up and running, and then the sizeable house and garden. His living room was set up as a perfect bachelor pad, dominated by a huge television, with a large stock of refrigerated beer. But, as we sat down for a drink on the veranda, and he took pot-shots at targets in the garden with his pistol, he revealed his concerns.

The problem was not that Jacques was ill-suited to the individualist, isolationist ideal *per se*: he was generally content with his own company. But he did want to get married, so as to have a companion in this choice of life. His father knew this too, and had suggested he begin to think about marriage now. The problem, said Jacques, was that he wanted a woman with education and ambition. Women who want to live with farmers, he complained, lack any real ambition in life, something he felt he had upset his mother about by telling her. His model for an appropriate partner presumably derived from his experience at Stellenbosch University. He now had his eye on a woman who worked as a conservation researcher in the game reserve attached to De Beers' nearby diamond mine. Here was a perfect combination: she had a goal, but it was compatible with life in the area. As his wife, she would have her own project. But Jacques had only met her once, and not in an overtly romantic capacity. She was the only option he could envisage at the time. There was simply no one else around who fitted the criteria. The masculine, individualist model of the pioneer left Jacques in a difficult position. His idea of an appropriate wife was a woman with her own work into which to throw herself. But he wanted to live a life centred on sole responsibility for a farm, his own project out of which he would make a success. In this model of work and its associated imaginary, a wife plays at best a supporting role. The role of work and the masculine individual in farmers' self-understandings had very palpable consequences for Jacques' chances of realising his more egalitarian notions of marital relations. This left him feeling increasingly worried about a possibly lonely future in his pioneering isolation.

The sons of farmers who are now reaching adulthood face different challenges from their elders. Koos ran the gauntlet of risk and sheer effort to establish Houmoed and then Grootplaas. But his long days of work lent themselves to romanticisation through pioneer stories, in which lone men hack back the bush. For his grandson, Jacques, there is no bush left to tame. He has taken over a run-down citrus operation, but the packshed and orchards are already there, and his father Willem keeps a protective eye on things. People in Jacques' generation struggle to create meaningful life projects. But the ideal of the individualistic pioneering male – still one to which Jacques aspires – confronts the reality of a landscape with little room to expand. It also confronts the fact that in this landscape relatively few other young men in his social group are challenged in the old way. Instead, a few big landowning families now dominate the border. The model of pioneering masculine individualism competes with contrasting expectations about sociality and marriage that derive from experiences at university and agricultural college.

Of course, in the greater scheme of things, Jacques is very lucky. His family's success has afforded him enormous privilege, including the opportunity to run a farm under his father's expert eye. Those who made a go of it alone and failed in border agriculture, or who never realised ambitions to acquire their own farm, feel far more sharply the burden of individualised notions of success. Understandably, they attempt to distance themselves from moralising interpretations of fortune and misfortune. Whereas I would often hear those who had left described as 'not serious' – suggesting laziness – Dirk took a different tone. Dirk, Jan's manager, is a Zimbabwean who had farmed on the border for twelve years before going bankrupt in the 1990s. Clearly his failure still smarts, all the more because the land he owned still bears his name (KhaDeki). And the distinction widely drawn in the area between hard-working success and its opposite cannot have helped. In Dirk's words:

Eventually it was impossible to go on. We had a few bad years and made a few bad decisions. It's one of those things. Nobody wants to stop farming. When your costs exceed your income, it doesn't work out, hey? A lot of us on this river at one time and a lot of people left, a lot of people went down the tubes here. Only a few people left [here] from the originals.

Dirk told me this in a quiet, somewhat resigned tone, with a lot of pauses between his words. When I said that it sounded to me as though he had lasted longer than most, he emphasised how long others had also managed to keep going, an understandably more forgiving view than the dominant one. But it was clear that failure had an enormous effect on him, from the way he related how he earned a living after losing the farm.

> I wandered around doing all sorts of things, *ja*, a lot of odd jobs. I had a garden service for a while which branched out into other things. *Ja*, it kept food on the table.... I even worked in a supermarket for a couple of years, *ja*, supermarket manager. As I say, I had my own garden service business. I did some building contract work. Just before I came here I was at [a safari lodge] for a year or so.

In effect, when he lost his farm, Dirk lost the sense of direction his project had given him. Successful farmers, in contrast, frame a future around continuing and expanding achievement. But such farms employ white managers as they expand. Conversely, managers contribute to the success of those who have achieved expansion. So not only did Dirk fail in his own enterprise; his work now contributes to another farmer's sense of individual success. The way Dirk speaks of his work as a manager lacks the progressivist sense of relentless improvement that farmer-landowners express. When I asked him what his job involved, he replied 'you tell me'. He was initially employed to look after the primary irrigation systems, the farm's 120 boreholes. But he subsequently also became responsible for the maintenance and picking of 300 hectares of citrus, maintaining the irrigation pivots over another 1000 hectares of cash crops including cotton, organising and supervising building work, and anything else that requires overseeing. 'I help out wherever,' he summarised. We will get a better sense of the key role played in the labour hierarchy by figures like Dirk – or André at Grootplaas – in Chapter 6. And it is worth emphasising that, as we shall see in that chapter, white managers' everyday work has most in common with the black foremen who coordinate the black workforce in and out of work time. What is key to note now is the low-status implication of Dirk's position as employed by a farmer, in an area with a dominant sense of success centred on tropes of ownership, self-employment, individual work, and initiative. This meaning was explicitly expressed by André, Grootplaas's white manager. He explained to me that his reputation among workers is that of a 'poor' person, in contrast with the wealth normally associated with white farmers. Permanent workers, he told me, joke that he is too poor to buy his own farm. Conversely, André valorises his proximity to and understanding of the workforce, by implicitly criticising those further up the ladder. He attributes Koos's younger relatives' lack of familiarity with TshiVenda to the family's snobbishness, born of extreme wealth. In white social circles on the border farms – including *braais* (barbecues) and polocrosse[10] tournaments – managers appear as relatively poor men, in comparison to local farmers who own aeroplanes.

[10] See footnote 7, above.

'It's in Our Blood, It's in Our Skin'

Even as farmers, their sons, and white managers contend with notions of success and failure, aspirations do not remain static. Farmers today have to adapt their goals – as enterprising individualists transforming their environments – to a new, highly uncertain reality. The final section of this chapter explains how farmers have used their projects to contend with an era of land claims and market liberalisation.

Success on the Move?

The Grootplaas view of the land as a canvas for development – for successful, ever-expanding enterprise – must be understood in the context of current uncertainties surrounding white agriculture. A focus on transformative work itself, rather than on landscape being transformed, reflects not only their self-understandings as 'serious' farmers, but also their response to post-apartheid land policies and to a liberalised, globalised agricultural market.

During the period of my fieldwork, Grootplaas and the other border farms were gazetted for land reform, something that had been expected for years. This was not an eviction notice, but rather an announcement that a group of black South Africans had claimed the land. The process must be pursued by the Restitution Commission, and if necessary go through the Land Claims Court. It will in all probability take many years, and the outcome is unknown. The claimants have to prove, under the terms of land restitution, that their ancestors were forcibly removed in the period since 1913.[11] In such a process, claimants and existing landowners mobilise experts to demonstrate or deny such displacement (see James 2007). The border farmers have commissioned a report by an anthropologist, which shows that their land has no history of settlement.[12] They contend that any signs of earlier habitation merely indicate the existence of riverside watering points for cattle. They argue that the area was too dry for villages, until the white farmers themselves installed large-scale irrigation systems and boreholes.

For the border farmers, the claim confirms a wider climate of uncertainty, created by land reform across South Africa, post-apartheid hostility to white farming, and the liberalisation of agriculture. Although they are contesting the claim, it underlines agriculture's precariousness. In response, the farmers have modified their existing ideals of success. Although farming on South Africa's northern border remains tied to a sense of transforming the landscape – of building something on the

---

[11] The year of the Land Act that first explicitly entrenched racialised land ownership in South Africa. See Introduction.
[12] The arguments against the claim were explained to me, but I was not allowed to see the report, which had been prepared as confidential evidence.

land – agriculture is coming to be viewed by its practitioners as a series of strategic business investments, rather than as a necessarily rooted way of life. Farmers adapt to their own risks – both international and national – by de-emphasising their long-term commitments, and emphasising their potential mobility.

This does not represent a sharp break with the past, however. For the border farmers, a flexible view of their enterprises corresponds to a pioneering ideal. Their arrival on the border was relatively recent, and was motivated by a desire to leave the places where they had farmed previously. Even since the 1980s, many of the farms have changed, as they have expanded and changed crops. Initially, most of the border estates had produced cotton for the domestic market. But landowners responded to liberalisation, international competition, and cuts in subsidies by turning to exports – their access to foreign exchange enabled them to weather the storm. In the 1990s, Koos, his two sons, and his son-in-law transformed their huge cotton plantations into citrus orchards.

Other farmers opted to diversify rather than change crop altogether, splitting their land between cotton, citrus, and sometimes vegetables. The mix of strategies is visible from the air, as the dark green rectangles of orchards jostle with round, pivot-irrigated fields along the Limpopo River. Grootplaas's neighbours have an especial advantage in cotton production: they own a gin, which enables them to process their crop and add value before they sell it, so cutting out a middleman and boosting profits. They now grow citrus for export alongside their cotton. Moreover, they can sell their ginning services to other farmers in the area, offering them yet another source of income.

Farmers have long relied on portfolios of investments to guard against an uncertain future. These are diverse, and the dangers themselves have changed over time. One established strategy to mitigate the vagaries of agriculture, I was told, is to run enterprises through separate umbrella holding and trading companies. These respectively own and manage a farmer's various portions of land along the border. The trading company leases the land from the holding company, so a bad crop scuppers the former, while leaving the latter and its estate intact. These family-run farms also use their companies to sidestep inheritance tax, since it is only the position of managing director that is passed down the generations. More recently, the threat of land claims gives the distinction between holding and trading companies new significance. The company that owns the estate and therefore participates in the land claim process does not own machinery or inputs, thus reportedly limiting the scope of negotiation. But, whatever the context, the explicit objective of such schemes is to 'spread the risk as far as possible', in the words of one farmer. Such sensitivity to risk has prepared the farmers well for recent uncertainty.

Indeed, some farmers' portfolios furnish them with options beyond farming on the border, an important asset today. One, for example, started a business packing fruit and vegetables in Messina (now Musina), the closest town, in 1987. As in the case of the cotton gin, above, this allows him to cut costs producing his own crops. But the business is now highly profitable in its own right, with 350 workers, and he spends a substantial proportion of his time in town, running it as a distinct enterprise. Subsequently, moreover, he has expanded to include a processing plant on the same premises (although it is a different company), drying fruit for domestic consumption and for export to the UK. These businesses offer him security as a source of income. In the context of land claims, they also offer a way to remain involved in agriculture and the farming community even if he loses his farm. Packsheds are central features of agricultural infrastructure, and are hubs in the organisation of farming areas. Plans such as these offer flexibility for farmers, as the border estates become options among others.

The Grootplaas farmers have been especially canny in providing for a flexible future. They began expanding their options with the purchase of land on the Mozambican border – Kleinplaas – a cold business calculation. The land was already claimed under South Africa's land reform programme, but they counted on an estimated ten-year delay on claims. Money could be made buying the estate cheaply, planting sugar cane (good for quick profit), and then selling at the government price were the claim to be successful.[13] Koos sent one of his sons, Rudi, to manage it, accompanied by a small number of senior workers. More recently, Grootplaas's owners leased land in Mozambique itself. There, according to Koos, civil war has left the country in an economic plight sufficiently severe to cause its government to welcome white farmers. As Koos sees it, Southern African countries will now cycle between welcoming white farmers when they are desperate and resenting them when they are prosperous.[14] For the youngest generation, yet further flexibility may be sought. As discussed above, Jacques was sent to Belgium after university to work the other end of the supply chain and acquire 'skills' (although he soon decided to return and farm).

On the border, there is a wide spectrum of farmers' strategies, but what unites them is that estates are seen in terms of shrewd business strategy rather than rooted settlement. The Grootplaas farmers have taken this principle even further than most, seeking investments well beyond the area, and even beyond South Africa, to mitigate risk. Most

---

[13] Like many such claims, it remains unresolved.
[14] Since the period of fieldwork, the farmers have abandoned this investment, not because of political difficulties, but because it was insufficiently profitable.

recently, the border farmers' hard-headed business acumen has led them to commission a coal-prospecting project on the farms: if they have to sell, they can ensure a higher price from an Australian mining corporation that has expressed interest than they would receive from the South African state following a land claim. By leasing back the orchards, they would at least have an income, along with the security of liquid capital. This is a way to remain flexible, but the strategy is itself risky. Farmers have already begun entering such agreements, but they do not yet know the impact of the new mine on air pollution. Coal dust may be detrimental to their crops. Yet the Grootplaas farmers have considered selling their land and trying their luck at leasing, for the right price – one that would recognise all improvements to the farm, even those that are irrelevant to mining. Doing so holds out a tantalising prospect: farming with minimal immobility. Ownership in this view comes to appear a handicap, rather than an asset. Indeed, one border farmer sees possible opportunity even in a successful land claim on his property, if his family is kept on to manage it with relatively free reign.

In such a climate, there are of course farmers who prefer to get out. Recently, Koos's son Paul has decided to sell his share in Grootplaas and buy a small guesthouse close by, for a quieter life. But most hope to stay in the sector. They see their estates in a much wider context, in which they hope to ensure the possibility of succeeding in agricultural enterprises *somewhere*, and are prepared to stay flexible to do so. The Grootplaas farmers' land in Mozambique was intended as a way out of South Africa altogether, if necessary, without leaving farming. Moreover, they were drawn to Mozambique by its relatively lax restrictions on moving wealth; in an uncertain future, the possibility of capital flight is attractive.

Farmers speak about these strategies in terms that are commensurate with their existing success stories. In their eyes they are now, as before, ensuring the conditions for transformative expansion. The amount of water locally available limits the extension of orchards and fields. The Grootplaas farmers have built a new dam to store water, much larger than the first. But buying Kleinplaas and leasing land across the border are ways to plough profit back into the business and avoid tax, since Grootplaas risked 'overcapitalisation'. As Castree (2009) argues of capitalism generally, Grootplaas's farmers are bound to a rhythm of reinvesting their profits as capital by expanding their enterprises over ever-greater stretches of land. Doing so also brings more farming opportunities for the family's male members. What appears a novel form of agricultural risk avoidance, therefore, can equally be understood as pioneering expansion under changing conditions.

What this produces on the border farms themselves is a strange tension. On the one hand, farmers consciously deracinate themselves from the estates they cultivate, even though these are their homes. On the other hand, the border farmers' broader views have remarkably little impact on their day-to-day operations. However flexible their plans, success depends in the end on deep commitment to particular projects, rooted in the land. As we have seen in this chapter, this is reflected in the ways farmers speak of their lives and their estates. It is also reflected in the choices they make about running their operations. Even as coal was being prospected, the Grootplaas farmers began construction of a vast second dam for irrigation, sinking their profits back into the farm despite current uncertainty about the future. Willem constantly establishes new arrangements with supermarkets and acquires funds to improve facilities in the labour compound. Farmers explain that, even in an uncertain environment, one has to keep expanding to survive: 'get big or get out'. As well as being a strategy to stay commercially competitive, this can also be seen as an assertion of permanence in the face of doubt. Changing the landscape is a way to make one's mark (see Hughes 2010 for similar responses in Zimbabwe). Border farmers today, therefore, think flexibly and spread risk, as twenty-first-century pioneers, while also being rooted in place by their personal investment in existing operations. Diversification vies with deepening commitment.

Conclusion

Pioneer stories reveal farmers' assumptions about race, class, and gender. They help us understand the racialised hierarchies on farms, at the pinnacle of which stand white, male farmers. But they also reveal how farmers see themselves in relation to their life projects. Taking this angle, we see white farmers less as static, stable 'types', and more as men with aspirations engaged in *becoming*. We also move beyond analysing *narratives* of pioneering self-sufficiency, to appreciate it as something that is also *practised*. Such self-sufficiency reflects the realities of farming; it is an object of aspiration among farmers that motivates practice; and it frames interpretations of successes, failures, and the pains of isolation.

Literature about wage labour and organised production is generally concerned to explore the plight of workers, but it rarely attempts to understand the ways employers understand their lives. This chapter has attempted to elucidate white farmers' perspectives about their life projects. It has examined how farmers consider themselves in relation to their work: their enterprises, their successes, and their failures. By focusing on the meaning of projects and of work for farmers, it has underlined how farmers view 'becoming a man' and making something of one's life. It employed this perspective to make sense of the recent tension in farmers' plans: both detaching themselves in the interests of flexibility, and continuing to invest deep personal commitment in particular agricultural projects.

The current uncertainties involved in agriculture do not affect farmers alone. We have seen in this chapter how farmers consider their success to be a matter of highly individual responsibility. However, focusing on farmers' self-representations, although shedding light on their own lives, gives an over-individualised view of commercial agriculture. In reality, the farms are home to large black working populations. Because farm workers' long-term precariousness is not reflected in the day-to-day dynamics of life at Grootplaas – the subject of much of the book – I leave direct consideration of their future prospects to the Conclusion. At this stage, it is important to note that the highly structured social arrangements that Grootplaas's residents develop, although they appear durable, depend fundamentally on the continued existence of the farm – something that cannot be guaranteed.

The chapter has focused attention on the border farms as working enterprises. This prepares the ground for the focus of the remainder of the book: how the farms as workplaces shape resident populations' lives. The next chapter steps back from the farmers. It examines the border farms as a recent example of a much longer history of border enterprise. Taking this view reveals the primary concern of border capitalism: establishing stable conditions for the control of territory and the recruitment of labour.

The issue with this is men talking to men...?

3 Behind the Mountain
Core, Periphery, and Control in the Limpopo Valley

Core, Periphery, and Passage

The previous chapter showed farmers' individualised views of their farms, in which whites exhibit self-sufficiency by establishing agricultural projects in remote places. However, these farms depend on huge workforces. To succeed, white farmers have to maintain a favourable environment for profit-generating production, which means ensuring consistent access to labour. Because workers live on-site, the farms are local social centres. From the perspective of black residents, they are spatial hubs of dense social webs that span the national boundary. The large, waged populations are also focal points of the border's various trade networks. This is an area that represents a coherent, meaningful unit in people's lives. From the point of view not only of farmers, but also of farm workers and other residents, the border estates organise people's different activities and relationships.

The Limpopo River's southern bank is the site of both large-scale agrarian capitalism and South Africa's border fence. The central role of labour-intensive workplaces makes the Zimbabwean-South African border different from many others. Anthropologists, keen to demonstrate that borders elsewhere are not the 'self-evident limits of the territory of states', have shown borders to connect, not just divide, populations (Pelkmans 2006: 12). Like the Lesotho-South African border, South Africa's boundary with Zimbabwe is 'itself a place – a unity created around a division' (Coplan 2001: 104). It is not, however, simply a 'border culture' (see Alvarez 1995; Donnan and Wilson 1999). The Zimbabwean-South African border does bisect an ethnic group's area of habitation,[1] that of the Venda. However, the stretch where border farming now operates is dry and inhospitable. Cross-border habitation occurred here largely because the farms themselves – agrarian capitalist enterprises – served as loci of employment and residence which drew people across. Much of the remainder of this book explores how people actually live their lives on the farms. What requires examining at this

---

[1] Like, for example, the Malawian-Mozambican case; see Englund 2002.

stage is how farmers and other border capitalists have shaped the area through their efforts to secure stable conditions for production.[2]

Border capitalists have had to assert control of territory and labour, on a major migration route and in the face of state policies that cast the area merely as a border to be policed. 'Because the people are moving through, in and out here, and we're the first set of farms on the border, we're obviously drawing a lot of attention.' Sitting on his terrace, his caged pet budgerigars nearby, the Limpopo River not far from his house, retired white farmer Jan described to me the effect on his family's cotton and citrus estate of being situated on the border. 'It is the place a person is going through,' he continued, 'so the local... authorities... are there to stop... infiltration of people... it's logical that they will concentrate on this area because this is where the people are moving.'[3]

The area lies squarely in the South African National Defence Force's ten-kilometre-deep militarised 'border zone'. According to this arrangement, the area is less a particular place – drawing people in from Zimbabwe and South Africa for work, trade, and other connections – and more an extended barrier on the edge of South Africa. There is a military 'echo' garrison every ten kilometres along the border fence, whose soldiers patrol the area daily. Soldiers operate throughout the zone, often staging ambushes on the dirt roads that run between the farms, to trap would-be migrants. Police face media and political pressure to show results in their fight against 'border jumpers'; they conduct raids to deport seasonal workers – many undocumented – who pick and pack produce.[4] Police and army efforts to find 'illegals' for the purpose of deportation challenge the arrangements enabling daily life on the border farms. 'They used to target my shop,' complained Jim, white partner in an estate and owner of the estate's general store:

[The police van would] stand there in front of the shop... and every customer had to show his ID document or his permit.... I went to them, and I said, 'Listen guys, I've got a bit of a problem with this. I'm not anti your campaign and your idea but you're actually affecting my business now.'

Farmers established their farms as capitalist enterprises and invested their sense of personal worth in their projects, as we saw in the previous

---

[2] The border farms discussed in this book are located to the west of Musina, stretching in the direction of Botswana. The history of border farming to the east, towards the Kruger National Park (and closer to the former homeland of Venda), differs in some respects – including an earlier reliance on South African labour tenants, and the later arrival of the current farmers. See Addison 2013.
[3] For Jan and others, see List of Key Characters at the front of this book.
[4] In 2009, the deportation of Zimbabweans was suspended. But, coinciding with the end of a programme to register undocumented Zimbabweans *en masse* (the Zimbabwe Documentation Project), deportations recommenced in late 2011. See footnote 22 of the introduction.

chapter. In so doing, they created hubs of the local economy. State officials' view of the area, on the other hand, is primarily as the border: the hyper-policed edge of a national territory. These opposed ways of thinking about the area have remained remarkably constant, despite obvious changes on both sides of the border.[5]

National boundary or local hub? These logics conflict. Border police and soldiers raid labour compounds in the middle of the night, and have difficulty distinguishing farm workers from transient 'border jumpers'. Bureaucratic delays in government departments mean that many seasonal workers are left undocumented, and they are often treated as 'illegals'. As Jan explained, the attention of border officials

is causing a lot of disturbance... because every time the police come and raid... they are disturbing the people that are legally here. Everybody's chased out of his house in the middle of the night and paraded all over the place, and he's got to show his ID documents and what-have-you and... then he's put into the van and is obviously taken first in for questioning and then released later... which is causing all the local legal people a lot of grief as well. Which we have always felt to be totally unnecessary, but if you look at it from a policing side, how else are they going to do it?... They've got to raid the living quarters where these people are. They've got to have a look at everybody's ID documents and make sure that they is [sic] legally here. Otherwise there'd be no control.

Belying this impression of effective state control, army and police ability to manage movement in the area is limited. The border is long and stretches of it are remote. The electric fence, erected during apartheid, has had its voltage turned down, and reportedly merely 'detects' people touching it. The signs of countless repairs expose how often it has been cut. Aggressive deportation raids on the farms are less a sign of authority than a response to its limits. They take advantage of easy and obvious targets – hundreds of workers concentrated in compounds – because police and military personnel find it so difficult to capture more transient Zimbabweans.

This chapter examines these opposing priorities, and the forms of control through which they are expressed locally: the Limpopo Valley as place of enterprise, versus the Limpopo Valley as border zone. Such efforts need understanding in terms of a wider history. Cross-border mobility, ineffective state attempts to control its boundaries, and local capitalism with resident workforces have all shaped the area over a long period. Indeed, when the border farmers initially came here in the midst of a border war, they were *both* part of border control and at the forefront of border capitalism. This chapter explores these enduring themes of

---

[5] Since 1994, white farmers in South Africa have had to negotiate with black authorities; since 2000, crisis in Zimbabwe has resulted in mass exodus, fuelling xenophobic sentiments and hysteria about 'border jumpers' south of the Limpopo.

South Africa's northern frontier. By so doing, it sets border farming in the context of the Limpopo Valley, long considered a border zone. A historical focus reveals border farming as a recent example of white attempts to establish large centres of production in the Limpopo Valley, and orient local dynamics around them. This meant creating places of sufficient economic importance to be considered a priority in their own right by state officials, and to be recognised by workers as a dependable source of employment. The border farmers, it will become clear, have largely succeeded in doing this, despite their claims to the contrary.

This account does not purport to be an exhaustive history of South Africa's northern border.[6] Instead, I focus on key moments that illuminate the uneasy coexistence of competing perspectives on the Limpopo Valley: those which foreground its position at the centre (a place that draws people in for work) and those emphasising its peripheral character. I begin by describing the development of border farming itself from the late 1970s. The existence of these farms resulted from government plans that underlined the Limpopo Valley's peripheral place in South Africa, a 'buffer zone' to shore up the apartheid state's territorial control. But they became a new centre of production – a place around which local lives became oriented.

I then widen the historical scope: first, showing how a distinctive border economy developed on the Limpopo; then examining the moments at which two important centres of production on the edge of a national territory were established. One was Messina Copper Mine. Founded at the turn of the twentieth century, this was key to the border's history – the first hub of wage labour, and the core around which Messina (later Musina) town came to be built. I focus on Messina's efforts to be recognised as an institution both with territorial jurisdiction on the border and with a monopoly over 'its' labour migrants. The other was the arrival of commercial crop farmers in the area north of the Soutpansberg from the 1930s. I explore the contestation that resulted between state employees who regarded the area as a peripheral zone, and agricultural capitalist developers who attempted to base their operations there and saw these as constituting the core of a hinterland from which they drew labour. Appreciating this struggle requires understanding state officials' priorities: they promoted territorial control as being of foremost importance,

---

[6] The northern border, with its extremely uneven regulation, was the site of different schemes for achieving a measure of control in the first half of the twentieth century. These were in turn the result of intricate and changing relations between different institutions – the Immigration Department, Native Affairs, the South African Police, and state-sanctioned labour recruiters (the latter themselves in shifting relationships with illicit recruiters). For in-depth and illuminating treatments of this story, see Jeeves 1983, and Andrew MacDonald's unpublished PhD thesis (2012).

while giving preference to the supply of workers to farms in the eastern Transvaal.

The chapter finally turns to recent attempts to resolve the tension between perspectives of centre and periphery, however provisionally. In the 1990s, the territory to the north of the Soutpansberg was declared a 'zone of exception', in which farmers had the special right to employ Zimbabweans. This formal exceptional status has since disappeared, but been replaced by *ad hoc* agreements between farmers and officials according to a very similar logic. Overall, the chapter reveals how the central logistical problem faced by employers – including the farmers described in the previous chapter – has been the control of labour in an area characterised by mobility and border policing.

Border Farms and Buffer Zone

Infiltration across the border has become a particular police and army priority in recent years. But despite a clear recent increase in the numbers of migrants, the border's economy has long revolved around similar southward movement. As one border farmer put it:

I don't know why there was an explosion of news in the media, because what they were complaining about ... happened for the last twenty years ... the people crossing the border.... They talked about a mass influx of people, and it's possible that there were a bit more people than normal, but it's a process that was going on for the last twenty years and all of a sudden everybody was talking about it.

Since its inception in the late 1970s, border agriculture has relied on such movement across the national boundary. In its early years, this reliance produced a tension between different understandings of the border. The white farmers came to the border in the first place as part of a state scheme to create a buffer zone in the border war of the late 1970s and 1980s. The apartheid government, plunged into a 'state of emergency' in its heartland, and now faced in addition with *Umkhonto we Sizwe* (Spear of the Nation)[7] incursions from the north, had very little territorial control along the Limpopo River. The farmers were brought to the area to enhance the state's capacity. However, they came to depend – for their Zimbabwean workers – on precisely the easy, unrestricted mobility across the Limpopo that they were intended to prevent. The new settlers both reinforced the border – which would soon be strengthened with an electric fence – and relied to some degree on its porosity.

The government attracted farmers with low-interest loans for land, and an electrical substation, telephone lines, and improved roads were

---

[7] The armed wing of the African National Congress during the anti-apartheid struggle.

constructed. In return, farmers were required to reside on their estates rather than being absentee landowners, and to participate in the local commando – building on earlier models of Boer militias in farming districts, the commando was part of South Africa's system of conscription for all adult white males. Willem, farmer at Grootplaas, explained:

> With the terrorists at that time that came from Zimbabwe, we were always on a standby, so they could call you anytime and say, 'Okay, well we've got a problem here.' ... We had a commander which would then phone us, or radio us, and say, 'Someone came through with landmines. ... We want everybody to get their kit and get together at a certain place and then follow their tracks and see whether we can pick them up.

This scheme created an entirely new kind of farming along the Limpopo River. Though surveyed and apportioned in the opening years of the twentieth century, the southern bank of the Limpopo River remained virtually uncultivated until the 1970s. White farmers had rarely been able to live on agriculture alone. Earlier, hard-up cattle farmers had worked on the nearby copper mine at Messina to supplement their incomes in bad years. Meanwhile, some successful white workers from Messina Mine would acquire land for part-time farming, sinking some of their income into what was in effect little more than an agrarian pastime. The land on which Jan's farm and its neighbour, Grootplaas, now lie was leased as Crown Land by the manager of Messina Mine in 1920,[8] bought by him in 1930,[9] used as a weekend retreat, and sold only in 1974 after his death.[10] Around this time, the large game and cattle estates along the river were bought and subdivided for profit, but still left mostly undeveloped. It was the arrival of a distinct new generation of young farmers – many Afrikaners, many from Rhodesia/Zimbabwe – that transformed the border area. They bought up the divided portions and irrigated intensively. As one of the older cattle farmers remembers:

> The government gave out these farms to farmers at the 4 per cent interest on the capital. They lent them the money to buy the farms at 4 per cent, 8 years without any interest. That's how all these people settled there. ... The whole river area, there was nothing! There were elephants!

The 'border war' both brought a degree of government support, in the form of 'soft' loans and infrastructure, and complicated the newcomers' primary aim of making a success of commercial farming. The border farmers along the river were focused on intensive irrigation for profit. Unlike the well-known agrarian capitalists of the eastern Transvaal – the

---

[8] Recorded in SAB URU. File details not included to maintain anonymity.
[9] Land Affairs, Pretoria: title deeds.
[10] Ibid.

'maize kings' (see Bradford 1993) – who often made their start-up capital in industry, some of these border cultivators had farming backgrounds. They came because they saw particular opportunity in the fertile alluvial soil of the Limpopo's southern bank. As we have already seen (Chapter 2), success on the land was important, a sign of individual worth and exhibited by profitability.

The insecurity of life on the border showed that South Africa's unwritten policy of deploying farmers as a buffer was not altogether effective. This was because of the situation on the South African side. What was conceived as a *buffer* against attacks by *Umkhonto we Sizwe* became a hub of residence and economic activity, and therefore a *site* – a target – for attacks. White residents carried radios, and children were driven to school in military convoys. Initially, as Thinus recalls, these new farmers were the border's military presence. 'We were supposed to be the security on the border. There were no army here. We were part of the army. Whilst farming, we were supposed to do the work the army and the police is doing today. That was part of the transaction.' As he remembers it, it was a spate of landmine attacks on the farms – around thirteen exploded, and more were planted – that brought a full-time army presence and the elaborate electric fence in the mid-1980s. The farmers' 'buffer zone' was also their home. As another farmer, Hendrik, recalls, 'it was a tough time. Our next-door neighbour, his house was attacked one night. And on this side, less than one kilometre from here, a landmine was detonated.' Farm work would begin each morning with landmine clearance, the manpower for which came from the black workforce.

In this context, participation in the commando came with advantages for farmers. They were issued with guns and ammunition, which they kept at home. And the commando became an important means of sociality among the young men who took up the loans. According to Hendrik, 'the circumstances, the conditions, actually forced us to stand together. We also had an obligation towards each other.' Willem elaborated on what this meant:

We had an attack on a farmer.... They shot a rifle grenade from the other side of the fence into the house. And so the next two or three days all the farmers got together and brought some of their builders... and we fixed up the house and put in new window frames and put on mesh wire.... And the women also had... a club... where they... did different projects to collect money to... spend on people that lost property... from attacks.

The commando drew the white community of the border together, not only when disaster struck, but also for training sessions, including for women: how to use rifles, target practice, and first aid.

Not that farmers presented this period as all-out warfare. Some were keen to moderate such an impression. One, for example, commented that 'it was stressful because you never knew what could happen next, but it was for a short time – about a year [he said 1987]. I think the people currently in South Africa that live in the cities has got a much, much harder time than we had those years, with all the crime going on.' Indeed, Jan, with whom we began the chapter, emphasised that he never felt forced to do anything except farm. For him, as for many along the border, the immediate comparison was with the Zimbabwean liberation war they had left behind. And for him, as for others, settlement along the river meant living peacefully, planting crops, and trying to make a success of his venture.

While the border's electric fence may have been erected in response to attacks on farmers, it became a barrier not only to *Umkhonto we Sizwe* guerrillas, but also to would-be farm workers. As Jan put it, 'in the '80s, the terrorist situation came. From then on, this distinct line was drawn.' The new agricultural estates prospered, however, because state control of the area remained limited, the fence was an imperfect barrier, and labour recruits found their way through. For farmers, policing the border, on the one hand, and establishing border enterprises whose hinterland lay across the national boundary, on the other, involved conflicting priorities. Over the years, they ceased to be concerned with border policing and were able to turn back to what they came for in the first place: building capitalist enterprises. But they continue to be affected by their location in a ten-kilometre-deep militarised zone on the edge of South Africa. As we shall see at the end of this chapter, border farmers have had considerable success persuading state officials of their estates' distinct roles as hubs in local patterns of settlement and movement. In doing so, they have used their location in the border zone as an advantage rather than just a hindrance.

First, however, I turn to demonstrate the even longer history of these dynamics. State efforts at control, migration, and attempts to orient local arrangements around labour-intensive enterprises lie at the heart of white capitalism on South Africa's northern frontier. Migration from countries to the north – especially Zimbabwe – has shaped the area since the Witwatersrand (the area around Johannesburg) began offering the prospect of cash-paid employment in the late nineteenth century. State officials have been concerned to police human movement in the Limpopo Valley, and been thwarted in their efforts, for just as long. And, while large-scale agrarian capitalism along the river itself emerged only in the 1970s and 1980s, white capitalists in the Limpopo Valley have pushed for state recognition for their enterprises as local centres since at least the 1910s. Understanding this historical pattern requires beginning with an analysis of how the area's economy was shaped by the nature – and limits – of early colonial settlement.

The Beginnings of a Border Labour Economy

The Limpopo Valley has an important history of pre-colonial settlement. Mapungubwe, the mountain-top centre of a state, abandoned around 1300 AD, lies not far from the river's southern bank near the point where South Africa, Zimbabwe, and Botswana now meet. And Musina town, 10 km shy of the Zimbabwean border today, was the site of intensive pre-colonial copper working until about the middle of the nineteenth century (see Mamadi 1942). But, by 1870, much of the area was '"a desolate region" beyond "where lions abounded and had driven out everybody".... Thereafter "not a human being could be discovered and the kraals... had long been deserted".'[11] This sense of remoteness was to remain a dominant perspective on the Limpopo Valley. In 1942 van Warmelo observed in a Department of Native Affairs ethnological publication, that 'the arid, flat mopani [sic] country between the Zoutpansberg and the Limpopo is, and has always been, virtually devoid of inhabitants' (1942: 4).

Early Boer settlement in the area was sparse. The Soutpansberg[12] mountain range had since 1848 been the site of the northernmost *voortrekker* town, Schoemansdal, about 10 km west of present-day Louis Trichardt/Makhado. Its economy had centred on hunting elephants and trading the ivory southwards. Attempts to subdue local African populations generated a trade in unfree child labourers, referred to as *inboekselinge* ('apprentices'). 'Rather than building towns, farms, and herds, the few burghers of the district invested their capital in hunting game and increased it by exporting trophies; children were taken as spoils of war because they, too, had export market value' (Boeyens 1994: 187; see also Delius and Trapido 1982).

Settlement in the northern part of Zoutpansberg District was characterised by repeated displacement. Boer settlement of the Limpopo region in the nineteenth century had been weak. But, 'if Zoutpansberg was for two decades a frontier, in 1867 it became an abandoned one, a casualty of the pioneering days when a white population of less than 30,000 spread itself so thinly north and east from the Vaal River as to overreach itself' (Wagner 1980: 316). Zoutpansberg was isolated in the north, 200 miles away from the rest of a South African Republic that 'was more properly an association of three distinct Boer communities, Potchefstroom, Lydenburg and Zoutpansberg, separated by nature and on occasion as profoundly by man' (ibid.: 318). Zoutpansbergers built their economy on the ivory trade, but recruiting black labour meant arming '*swart skuts*' (literally 'black marksmen'), who eventually refused to return their guns.

---

[11] In Lieutenant Frederick Elton's 1872 account of his attempt to navigate the Limpopo, in Bonner 2003a: 30–1.
[12] Modern spelling.

Ultimately, this led to the Venda driving the Boers out (ibid.). With their expulsion ended the trade in child *inboekselinge*. But coordinated efforts to export people from the Zoutpansberg to areas further south as labour were to characterise the area well into the twentieth century. The transfer of co-opted labour migrants from the north to areas further south later became a focus of disputes over the Limpopo Valley's relationship to the rest of South Africa.

Displacement, and mobility more generally, made the area an important thoroughfare to and from the region to the north of what later became the border. But this in turn necessitated greater 'stabilisation'. The Limpopo River was – and remains – traversable on foot for much of each year, and the surrounding populations are 'in language and tradition a mixture . . . the result of successive migrations from the north and west' (Fouché 1937, in Bonner 2003d: 11). The settlement of Rhodesia by British and Afrikaner farmers in the 1890s added more people to those already traversing the valley, with Pont's Drift and Rhodes' Drift connecting the new colony to its southern neighbour across the Limpopo River. The increased expansion of Boer settlement northwards then brought further mobility. Black groups in the valley resisted new Transvaal laws prescribing black 'locations' (reserves) and imposing taxes. One of these groups was the western Venda, who had migrated from north of the Limpopo a century earlier. In the midst of a succession dispute, Chief Mphephu was forced across the river after defeat in 1898 by a large Boer commando dispatched to remove him from the area (see Braun 2013).

The Boer victory, and subsequent plans to open up land for settlement, saw white squatters move into the mountains. These incursions by poor settlers, contravening official proscriptions, coincided with 'the return in "great numbers" of dislocated Venda farmers'. The result was 'administrative chaos that lasted for decades' (ibid.: 279).[13] At the same time, members of the Boer commando were to be paid through the system of 'Occupatie Wet' (Occupation Law), 'which exchanged land – most of which was neither under its [the Republic's] control nor surveyed – in return for residence and military service. . . . Its simple purpose was to stabilise the area by creating a white security buffer among the previously intractable African chiefdoms' (Bonner 2003b: 32–3). This notion of a white buffer in the north was a precursor of attempts to create such a buffer in the 1980s, as discussed in the previous section of this chapter.

White occupation of the Limpopo Valley remained irregular. The 1896–7 rinderpest destroyed over 90 per cent of the subcontinent's cattle. Absentee ownership and land speculation were rife, covering the best part of the land for which title deeds had been issued (ibid.: 33). And the Second South African War (1899–1902) created an opportunity for

---

[13] It was also at this time that the town of Louis Trichardt was erected (Braun 2013: 279).

displaced black populations to reclaim land when British forces enacted punitive counter-insurgency techniques against Boers in the Zoutpansberg region. The Venda chief, Mphephu, returned to the south side of the Limpopo after the war. British officials, concerned about conflict resulting from his attempts to reassert control over territory, sent Mphephu back north again. But it soon became clear that their hold over the region was not strong enough to govern without him (Braun 2013). Disarmed, he was relocated east, to what later became the Venda homeland (Motenda 1942: 61).

By the turn of the twentieth century, South Africa's northern frontier was characterised by extremely sparse settlement, but also by constant movement and cultural hybridity:

The nineteenth- and early twentieth-century residents of the Mapungubwe area comprised Venda, Twanamba, Kalanga, Birwa, Lemba and Tlokwa alongside an occasional white farmer. Their scattered settlements in this area represent historical sediments deposited there by larger processes of migration, conquest and trade. (Bonner 2003d: 11)

State presence in the Limpopo Valley of the early twentieth century was practically non-existent. Zoutpansberg District, the northernmost part of the Transvaal, was divided by the Soutpansberg Mountains (see Map 2). Sibasa, Spelonken, and Pietersburg, the District's administrative centres, all lay to the south of the range. The border with Southern Rhodesia lay to the north, in the Limpopo Valley, a dry, flat expanse of *mopaneveld*. Although the proportion of Mphephu's people living as labour tenants on white farms may have been as high as 70 per cent by 1907 (Braun 2013: 290), very few white people lived north of the mountains. Indeed, from the perspective of white officials, it remained a barely liveable backwater. In 1904, as one official put it, 'a very large extent of uninhabited country lies between the limit occupied by our Natives and the Limpopo.... Owing to the wild and waterless nature of this country, only recognised paths [between isolated kraals] are used in travelling to Rhodesia.'[14]

It was in this context, lacking personnel or direct control near the border, that state officials addressed two related challenges to their authority in the Limpopo Valley: the continuous movement of non-South Africans southwards, and the illicit labour 'bandits' whose business it was to recruit them, voluntarily or otherwise. Central to state officials' preoccupations, then, were attempts to control the movement and recruitment of migrants.

As van Onselen notes, understanding migration southwards into South Africa in the early twentieth century requires seeing the whole region,

[14] TAB SNA 191 NA4/04, 10 April 1904.

including Southern and Northern Rhodesia, as part of the same economic system. Within it,

> The essential objective of the African worker was two-fold: first, he aimed to avoid as far as was humanly possible the prospect of work in the zones of low wages; second, he sought to move as far south as possible. The further south he got, the higher the wages he was likely to receive. Ideally, the goal for a black worker was to sell his labour power in Johannesburg. (1976: 228–9)

South African work destinations were generally more attractive than Southern Rhodesian ones, and men worked their way south to reach the Limpopo. Having crossed it, the same gradually improving work conditions applied within the Transvaal, although Messina Mine on the border was relatively desirable (ibid.: 318, endnote 4). Increasing numbers of migrants from ever-further parts of Southern Rhodesia crossed the Limpopo in the twentieth century: by the mid-twenties, 'truckloads of black workers made their way south to the border area around Beitbridge' (ibid.: 237). The greater resources of gold mines on the Witwatersrand continued to make it the ideal destination. Other workplaces like Messina – and later the Limpopo Valley farms – might be staging posts en route, for workers seeking not only pay but also nodes in regional information networks (ibid.).

It was virtually impossible for the South African authorities to prevent this migration pattern. But the question of a lack of state control north of the Soutpansberg tended to focus on the equally intractable problem of illicit recruiting agents – known as 'touts', 'bandits', or 'pirates' – who picked off migrants as they headed south. A 1913 report prepared for the Native Commissioner at Pietersburg gives a sense of how 'labour pirates' were officially viewed:

> The manner, in which recruiting is done across the borders, amounts to little short of highway robbery. A [licensed] runner with a gang of boys is waylaid and the boys taken from him whether they are willing or not. Of course, the latter do not care who gets the capitation fee for them as long as they can get down to work... The local term for [the illegal agents] as a class, is the 'Bandits' and I do not regard it as too severe a one.... They are all at loggerheads one with the other, and, but for their cowardly attributes which undoubtedly have become accentuated by the lives they lead, violent disruptions would most assuredly occur.[15]

Despite changing state policy regarding the recruitment of migrants from the north (see below), white outcasts of various kinds continued illegally to divert border crossers and supply them to employers in South Africa

---

[15] SAB NTS 2025 26/280 2598/14/473. 'Extract from report submitted to the Native Commissioner Pietersburg on the 13th August 1913 No. 56/13/405'.

for a fee. Indeed, illicit recruitment was often the means by which licensed agents acquired their workers. 'Shopkeepers, traders, ivory poachers, and big game hunters often doubled as informal brokers acting on behalf of licensed recruiters' (Murray 1995: 385). 'Labour pirates' were a characteristic of the northern border until at least the mid-1940s, when 'state officials at Messina on the border complained bitterly that "the recruitment of foreign natives is a racket out of hand"' (ibid.: 396). As Murray argues, their existence complicates sharp conceptual distinctions between free and unfree labour during this period. They connected labour supply (migrants) to demand (mines and farms), but sometimes did this by recruiting at gunpoint. Such coercion even characterised the supply of labour to the Witwatersrand goldfields (ibid.: 377). The shadowy operations of 'labour pirates' not only threatened state assertions of territorial control, but also challenged local claims to passing migrants as large white enterprises grew up in the Limpopo Valley in the first half of the twentieth century.

By the early twentieth century, then, battle lines around control of territory and human mobility in the Limpopo Valley had been drawn. South African authorities had a precarious grip on local goings on. Migrants' attempts to head south for work at decent wages were challenged by labour touts' attempts to divert them to various employers and licensed labour agents for a fee. A new player entered the fray as border capitalists made efforts to cast the area as a distinct local economic centre, both asserting spatial control and claiming migrants as their 'natural' – therefore rightful – labour supply.

A Centre at the Periphery: Land and Labour in a Company Town, 1910–20

Messina Mine tamed the Transvaal's northern frontier, and became an economic hub in the Limpopo Valley. It would attract a large migrant workforce and grow into the border's major town. From the perspective of the few local white cattle farmers, it would represent both a place of potential employment and a source of urbanites seeking rural recreation. In time, it would shape its surroundings profoundly. The year 1904 therefore marked a key turning point for the border region. In March of that year, the British Lieutenant Colonel John Pascoe Grenfell obtained a Discoverer's Certificate for the copper a little south of the Limpopo River's southern bank, 'on certain unproclaimed Government Ground situated on the farm BERKENRODE 1424, 1425' in the Transvaal Colony (certificate quoted in full in Mills 1952: 9).

Grenfell had heard rumours of a history of African copper mining in the area while in Rhodesia during the Second South African War.

Figure 3.1. Messina mining camp in 1907. (Photo courtesy of Brian Kalshoven.)

He interested two Imperial soldiers in the idea, and in 1901 visited the proposed mining area. At the end of 1902, once the war had ceased, a prospecting team set out northward from Pietersburg, with 'two wagons, 32 oxen, arms and ammunition, food for several months, and six or seven horses... from an army remount depot' (Messina (Transvaal) Development Company 1954a: 5). The mission produced considerable excitement. As the surveyor, Webber, wrote in a report to Grenfell: 'We were dumbfounded! The whole farm is a mass of copper' (quoted in Messina Publicity Association c. 1961: 3). Writing to his brother in August 1903, Grenfell recounted Webber's observations, calling the find 'a veritable Copper Rand' (MTDC 1954a: 5). Despite Grenfell's official status as 'discoverer', copper mining had a long history in the Musina area. Accounts differ as to whether the Musina copper miners migrated from Phalaborwa to the southeast (Mamadi 1942), whether they were Lemba from the north (Stayt and Thompson, cited in Bonner 2003c: 38), or whether the groups overlapped or coexisted (ibid.). Either way, it is clear that copper working had been practised on a large scale. Precolonial mining had ceased in the nineteenth century, when Venda and Sotho raiders scattered the Musina community (Mamadi 1942: 83–4).

Grenfell's Messina Mine created the first concentrated white settlement north of the Soutpansberg (see Figures 3.1 and 3.2). In a somewhat

Figure 3.2. Early shaft sinking at Messina. (Photo courtesy of Brian Kalshoven.)

triumphalist tone, a fiftieth-anniversary mine publication recounts the early days, once financial backing was secured from London:

In the bak-veld, the pioneers sank their shafts. They took on native labourers. Dosing themselves with quinine against malaria they worked on, widening the overgrown trails and building mud and straw huts. The mine and the new community were conquering the scrub. (MTDC 1954a: 6)

Another 1954 company pamphlet trumpeted a half-century of expansion:

Now, employing 500 Europeans and 4,500 Natives, it can be claimed that Company has attained the stature of a large producing company with its own niche in the financial structure of the country, being the only company to produce refined copper in the Union of South Africa, and it enters the 50th year of its existence with renewed confidence for the future. (MTDC 1954b: 14)

To these authors, Messina had not only succeeded as a company, but also 'conquered the scrub'. In 1907, the completion of a road from Pietersburg brought a bi-weekly coach service, and 1908 saw the beginnings of expansion onto new land. Soon after, capital was raised for a concentrating mill and smelting furnace (MTDC 1954b) (see Figure 3.3). In 1914, a railway link south was completed for the hand-picked ore. Residential areas were developed. By 1954, 'with the introduction during the last few years of parks, lawns and flower gardens alongside the tarred thoroughfares, the building of the new Hospital, Company Offices, Recreation Club, and numerous attractive houses', Messina had been 'changed and improved beyond compare' (ibid.: 18).

Figure 3.3. The new mill, completed in 1913. (Photo courtesy of Brian Kalshoven.)

Despite the apparent inevitability of Messina's success from a 1950s vantage point, the mine had been vulnerable. Fluctuating copper prices caused regular crises (temporary closure until government intervened in 1922; crisis until further intervention in the 1930s [MTDC 1954a: 12–13]; eventual closure, following the 1980s crash, in the early 1990s). Added to these economic problems were two key difficulties Messina faced as it grew. Management had to struggle to establish territorial government that went beyond frontier policing. And it needed to secure labour. This was no easy task, despite – or in part because of – the Limpopo Valley's saturation with labour touts.

Controlling Territory

Messina's location – the Limpopo Valley – had been a place for the South African authorities to police (with little success), rather than govern. Now, Messina's managers needed to establish it as a place in its own right. The mine management approached the problem by focusing on health. In 1914, the Company discussed establishing a Health Committee to combat malaria and other illnesses. A malaria outbreak the following April lent the proposal a sense of urgency,[16] no doubt confirmed by

---

[16] TAB TPB 1025 TALG 7683. 'Messina Health Committee Establishment'. Provincial Secretary to Magistrate, Louis Trichardt, 4 January 1915; Executive Committee Resolution, 28 January 1915; Magistrate, Louis Trichardt, to Provincial Secretary, 24 April 1915.

smallpox on the mine later in the year.[17] Negotiations on the scope of the Committee's authority soon focused on its territorial remit. How much territorial control should the Health Committee have, given its affiliation to – indeed, its part in – the Messina (Transvaal) Development Company, but its stated aim to provide a broader public good? The Committee's answer to this was clear:

> The Messina Health Committee has the honour to apply for jurisdiction over a larger area than what has been granted to it by the Government.... Perhaps the principal object of the Health Committee is to prevent malaria, and to do this, it is necessary to control a large district, in order to prevent the interchange of infection between different parts of the district brought about by the movement of people.[18]

The government took this plan (see Figure 3.4) under consideration, but local absentee landowners were strongly opposed. At the heart of their objection was the attempted control by private businessmen over matters of administration that ought to be the remit of the state. The Transvaal Consolidated Land and Exploration Company complained to the Provincial Secretary about being left out of the loop, despite the fact that 'this area, which is equal to a fairly large English county, includes a number of farms which belong to my Company'.[19] They objected that the area was too big to administer properly in the first place. One of the Health Committee's goals was to limit the residence of 'squatters' – black people living around Messina without mine employment – to controlled areas. Transvaal Consolidated noted wryly – and with a hint of sarcasm – that it 'might for this purpose have applied, with as much reason, for jurisdiction over the whole of the Zoutpansberg District as over the area in question'.[20]

The problem was more general than controlling the residence of squatters. This goal was not itself contested by other landowners. However, despite Messina's assurances that they were acting in the public interest, they appeared to be instituting a form of territorial government. Oceana, a company with intentions to begin mining copper, was being asked to place its land under the direct jurisdiction of a body representing the Messina Company. It did not wish to be bound by the decisions of a

---

[17] TAB GNLB 43 984/12/41. 'Rhodesian Natives employed on Messina Copper Mine'. Mine Secretary, Messina, to Director of Native Labour, 18 October 1915.

[18] TAB TPB 1025 TALG 7683. 'Messina Health Committee Establishment'. Messina (Transvaal) Development Company to Resident Magistrate, Louis Trichardt, 17 August 1915.

[19] TAB TPB 1025 TALG 7683. 'Messina Health Committee Establishment'. Transvaal Consolidated Land and Exploration Company to Provincial Secretary, 6 September 1915.

[20] TAB TPB 1025 TALG 7683. 'Messina Health Committee Establishment'. Transvaal Consolidated Land and Exploration Company to Provincial Secretary, 11 September 1915.

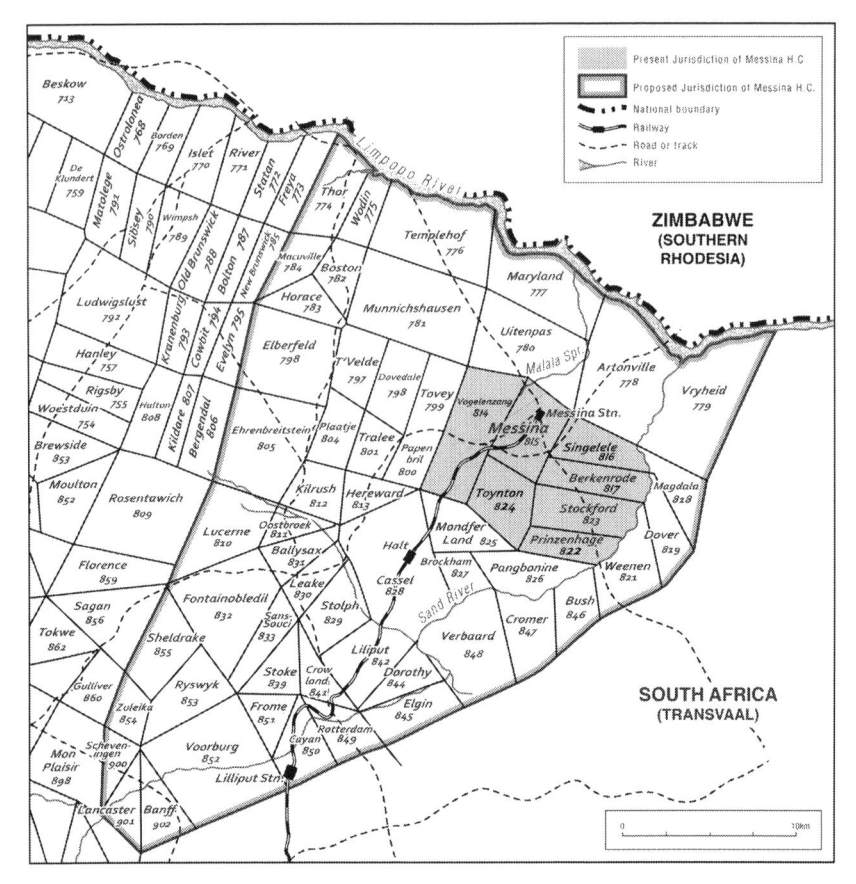

Figure 3.4. Planned expansion of Messina Health Committee's jurisdiction, with existing jurisdiction shaded. (Based on TPB 1025 TALG 7683, 'Messina Health Committee Establishment', country/province names added for clarity.)

competitor, including the clearing away of the very timber that would be essential for building its own infrastructure.[21] The representative of another company, Henderson Consolidated, 'view[ed] with grave suspicion such a secret and mysterious attempt to obtain control of my Company's properties, where any straightforward and honest request for facilities for any furtherance of public health would have been so readily considered on its merits'.[22]

---

[21] TAB TPB 1025 TALG 7683. 'Messina Health Committee Establishment'. Oceana Consolidated Company to Provincial Secretary, 6 September 1915.
[22] TAB TPB 1025 TALG 7683. 'Messina Health Committee Establishment'. Henderson Consolidated Corporation to Provincial Secretary, 9 September 1915.

Local landowning companies were concerned that Messina Health Committee was attempting to establish itself as a governing institution, but with an interest in ensuring advantage for the Company. Henderson Consolidated put it in the strongest words:

You will note that they [the Committee members] are all practically under the control of Mr Emery, the Manager of Messina Company, and I consider that it is for certain undisclosed and entirely unjustified objects, connected with the running of his mine, that this gentleman has in view in making the application he has made.[23]

Faced with such opposition, Messina backed off. The Health Committee gained jurisdiction over the 'reduced area' over which the mine already had some control (the area shaded in Figure 3.4 above). It quickly extended its authority beyond sanitation, becoming the town's government, and was replaced by regular municipal structures only in 1968.

Controlling Labour

There was a further reason to establish Messina as a recognised local centre, with control over human mobility and settlement. Messina created an unprecedented demand for workers on South Africa's northern border, and it did so on the route south to the Witwatersrand. Unlike white enterprises in earlier periods, Messina Mine's survival depended on large numbers of men *staying* in the area to work, rather than simply passing through. Expansion, from mine camp to mine town to municipality, depended on controlling *territory*. But Messina's success as a productive enterprise depended on orienting people and their movements towards the mine – controlling *labour*.

In March 1912, the Local Board of the Messina (Transvaal) Development Company contacted the Director of Native Affairs, in something of a panic.[24] The company had an agreement with the government regarding the railway connection soon to be completed, and was attempting to prepare for the connection by developing the mine extensively. Its labour supply was falling 50 per cent short. Messina's location made it difficult to guarantee a workforce, since the area was only sparsely populated. The Board explained that 'there are no large kraals within a good many miles of Messina'. Indeed, they claimed, 'there are only several isolated huts within this area, and the few natives residing there do not as a rule seek work'.

---

[23] TAB TPB 1025 TALG 7683. 'Messina Health Committee Establishment'. Henderson Consolidated Corporation to Provincial Secretary, 28 September 1915.

[24] TAB GNLB 43 984/12. 'Suggestion that the District around Messina Mine be closed to recruiting.' Acting Secretary to Local Board, Messina, to Director of Native Affairs, Johannesburg, 13 March 1912.

Attempts to recruit more widely were being thwarted. Labour supply came from two sources. Some hailed from Spelonken (see Map 2), already so 'overrun' with recruiters for mines on the Witwatersrand 'that it may be taken for granted that boys who come from there to Messina are old employees and cannot be recruited or persuaded to go to the Rand, or elsewhere'. The rest – the majority – were southward migrants from Rhodesia. By August 1915, around 1,400 black workers out of a total of 2,039 came from north of the Limpopo River, and it was assumed by the mine's general manager and by the government's Acting Director of Native Labour that 'a large percentage' were Rhodesian.[25]

White recruiters and their 'native runners' were stealing potential workers from under Messina's nose. 'The shortage of labour at Messina', complained the company's board, 'is largely due to the increased activity of Native Labour Agents and we are continually faced with difficulties to counteract their efforts to prevent natives from proceeding to Messina, and also taking natives from our Compound already in our employ'. A white agent would 'usually den[y] being at Messina for the purposes of obtaining labour and gives as an excuse that he is hunting or fishing'. The fact that potential recruits crossing the border followed particular paths as they made their way towards Messina made it easy to find them. 'Boys who express willingness to engage for the Rand are diverted past Messina and others are persuaded by the usual means to proceed south and are kept clear of Messina.'[26] Messina's Board asserted that their claim to the labour of migrants passing through their territory should supersede the aims of migrants to head southwards. The Limpopo Valley should be regarded as Messina's proper catchment area, rather than as simply the northern frontier of South Africa.

The protection Messina sought from the Director of Native Affairs asserted their territorial claim, through a ban on recruitment near the mine: 'an area equal to a radius of say 25 miles from Messina should be closed against Native Labour Agents and Runners'. The request was refused, on the perhaps spurious grounds that so many similar requests had been received from employers, for protection of their perceived 'natural labour reserves'.[27] This, understandably, provoked bitter disappointment on the part of Messina's Board that the government 'do not offer

---

[25] SAB NTS 2025 26/280 2598/14/473 DNL.1950/13/D240. 'Illicit recruiting Native labour in Southern Rhodesia.' Acting Director of Native Labour to Secretary for Native Affairs, 27 August 1915.

[26] TAB GNLB 43 984/12. 'Suggestion that the District around Messina Mine be closed to recruiting.' Acting Secretary to Local Board, Messina, to Director of Native Affairs, Johannesburg, 13 March 1912.

[27] TAB GNLB 43 984/12/3269/12. Director of Native Labour, Johannesburg, to Acting Secretary to Local Board, Messina, 12 April 1912.

us any assistance', especially since they were aware that allowances *had* been made in other, similar cases.[28]

This was not the first time that control of labour had been raised as an issue for Messina. The Native Labour Bureau had, in fact, already considered proclaiming Messina a Labour District in 1911, to protect it from labour agents. But it had decided the territory that could reasonably be claimed by the mine under such terms would not reach the Limpopo River, thus defeating the purpose.[29] In any case, proclaiming a Labour District would have come with the downside for the mine of increased government attention. This was in fact suggested by the Secretary of Native Affairs following the alleged ill-treatment of black workers: 'the Manager [of Messina] might be informed that if it should be necessary to proclaim a Labour District in order to protect the employees of the Messina Mine, the Compound Manager would require a licence... the issue and renewal of which is under strict scrutiny'.[30] Formalising a relationship with government by creating a territorialised Labour District came not just with a monopoly over recruits in the area, but also with heightened bureaucratic surveillance. Messina, however, wanted it both ways: territorial control both to manage their own affairs and to ensure a labour supply.

Trouble with recruiting did not disappear. Shortage of labour was often due to desertion, linked to poor conditions and workers' aims to reach better employment (see van Onselen 1976). In 1915, the visits of 'a clique of Labour Agents recruiting for the Rand... coincided with the disappearance of the Natives, who cleared off without notice and without finishing their month's work'.[31] This was virtually impossible to prevent, since 'there are so many ways of getting through the bush to the South West of Messina, that there is no possibility of getting conclusive results either by Police patrols or by restricting the issue of Passes'.

Part of the problem was noted in a government report on conditions at the mine. Infrastructure and accommodation were extremely basic: the hospital was a single hut; compound living quarters were overcrowded; huts in the location for married workers were 'of a very poor type generally whilst old sacks and paraffin tins play a great part in protection against the elements'; and latrines were 'obviously inadequate seeing as there are

---

[28] TAB GNLB 43 984/12. Acting Secretary to Local Board, Messina, to Director of Native Labour, Johannesburg, 23 April 1912.
[29] TAB GNLB 43 Licences 1285/11/2879/11. Acting Asst. Director, Govt Native Labour Bureau, to Acting Secretary for Native Affairs, 13 September 1911.
[30] TAB GNLB 43 984/12/D154. 'Treatment of Native: Messina Copper Mine'. Secretary of Native Affairs to Native Commissioner, Louis Trichardt, 29 September 1914.
[31] TAB GNLB 43 984/12/53. LT No. 33/15/1079. 'Messina Copper Mines', 1 December 1915.

only two latrines containing 8 buckets in each for a complement of rather over 2,000 Natives'.[32] Even given that mine compounds elsewhere were often unpleasant, such living conditions make it understandable that workers would leave if offered employment on the gold mines, where bigger mine companies had better-established facilities and could pay more.

By 1918, the mine's dual concerns of territorial control and ensuring a labour supply had been combined. The recruitment of migrants from north of latitude 22°S was, in theory, prohibited.[33] This should have helped Messina, which was still allowed to employ Rhodesians. But the South African authorities, lacking capacity on the northern border, allowed recruiters to continue their operations. In late 1920, the Messina Company wrote again to the Director of Native Labour: 'if restrictions could be placed upon the recruiting of natives within the area controlled by the Messina Health Committee it would be of considerable assistance'.[34] By the end of the year, the area of the Messina Health Committee's jurisdiction – the reduced territory of seven farms (see Figure 3.4 above) – was a specifically prohibited area for labour recruiters.[35] Even then, however, a lack of state capacity placed Messina at the mercy of 'labour pirates'. As late as the mid-1940s, Messina would continue to complain about the 'racket' in clandestine recruitment (Murray 1995: 396).

The establishment of Messina Mine reveals the meanings of territorial control for a company that was hungry for workers. Messina faced two problems in the years examined here. Both involved attempting to establish the area as a centre, a place of economic importance. The first problem concerned proper government on the frontier. Attention to the Limpopo Valley by South African state officials was about policing, but Messina needed to establish northern Zoutpansberg as a place in its own right, not just the frontier of the larger entity of South Africa. The second problem was about securing labour for the mine, which also took territorial form, as Messina claimed monopoly rights to recruit workers within its area of jurisdiction.

The South African state was to continue to struggle with labour recruiters. Strategies to control 'clandestine immigration' and illicit recruitment

---

[32] TAB GNLB 43 984/12/53. 'Messina Copper Mines', August 1915.
[33] Government Notice No. 1693 of 1918 banned the recruitment of migrants from north of latitude 22°S (except with special permission). This built on earlier restrictions (see Jeeves 1983), whose rationale was the high mortality of so-called 'tropical natives' on South African mines.
[34] TAB GNLB 43 984/12/53. 'Messina Copper Mines'. Messina (Transvaal) Development Company to Director of Native Labour, 19 November 1920.
[35] TAB GNLB 43 984/12/53. 'Messina Copper Mines'. Messina (Transvaal) Development Company to Director of Native Labour, 24 December 1920.

Behind the Mountain

in the 1920s, 1930s, and 1940s reveal that government officials continued to see the Limpopo Valley primarily as a borderland. But, as the border region transformed in the wake of Messina's consolidation, a growing community of commercial crop farmers emerged to the north of the Soutpansberg. The mine's predicaments offer insights into the difficulties that these cultivators would face. And, indeed, as farmers strove to establish places of production, they would come to reiterate Messina's claim that the area was not merely a thoroughfare southwards.

Supply, Demand, and Territory: 'Clandestine Immigration' and Farm Labour

Following the period just described, there were significant continuities in South African policy towards Rhodesian migrants between around 1918 and 1950: one was the abiding dilemma about what to do with a northern border that was very hard to police effectively. Another continuity was that Rhodesians came to be regarded as fodder in officials' compliant responses to the demands of big Transvaal agricultural producers. For these farmers, foreign migrant labour was crucial, because of perennial difficulties with finding workers of local provenance. These continuities endured, despite shifting alliances between farmers and national political parties, as well as important differences between administrations in the South African government (see Bradford 1993). My focus here is on the continuities themselves. Policies regarding 'clandestine immigration' during this period reveal the continuing relationship between governing *territory* – and therefore the movements of people – and supplying *labour* to white capitalists.

The Rise of Limpopo Commercial Agriculture and the Louis Trichardt Labour Depot

In 1948, the Zoutpansberg Farmers' Union wrote to the Native Commissioner, Louis Trichardt, describing the developing Limpopo Valley crop farms. 'Tens of thousands of pounds and much pioneering work have been put onto the farms along the Njelele River during the past fifteen years, and this industry has been built up entirely by prohibited immigrant labour.'[36]

According to one longstanding resident of the valley – a doctor and cattle farmer born in 1926 – the 1930s saw an increase in the number of farmers, but most put their efforts into cattle. Nevertheless, a particular

---

[36] TAB HKN 1/1/52 17N3/20/2, 'Foreign Farm Labour Scheme Louis Trichardt Main File.' Charles Chamberlain, Soutpansberg Farmers' Union to Native Commissioner, Louis Trichardt, 25 February 1948.

pocket of land between the Soutpansberg and Messina (see Map 2) had been the site of nascent agrarian capitalism since the 1930s. A new government-funded dam was completed on the Njelele (in TshiVenda: *Nzhelele*) River early in 1948, enabling much greater expansion. By the early 1960s, a Messina promotional publication could declare that

> The old arid area of the Soutpansberg is being rapidly transformed into the country's winter larder, where grapes, vegetables, citrus and sub-tropical fruits flourish. The quality of the citrus is particularly high and a ready export market is found. 114,000 tons of fruit and 9,000 gallons of cream are railed annually to the markets of Pretoria and the Witwatersrand. (Messina Publicity Association c. 1961: 4)

Like Messina Mine before it, the Njelele farming area was highly dependent on migrants passing southwards through the farms. In 1947, the Chipise[37] *Boerevereniging* (farmers' union) complained to the Native Commissioner, Louis Trichardt, about the agents recruiting in the area for farmers elsewhere.[38] As the Native Commissioner informed the Chief Native Commissioner at Pietersburg in 1948, 'the foreign Natives entering this district from the North and presenting themselves at the Njelele farms on their way southwards have hitherto been the source of the labour employed by the Njelele farmers'.[39]

It was in 1948, too, that this dependence on northern migrants brought the farmers into conflict with the authorities. In the late 1940s, the northern border region was the site of new labour depots designed to channel Rhodesians crossing the border to farmers, most of whom were situated outside the Njelele region. A network of such depots was established in March 1947, centred on Louis Trichardt. They were intended to 'serve as collecting points from where the recruits will be conveyed by motor lorry to the main depot at Louis Trichardt'.[40] Migrants would be brought to these subsidiary depots by 'native' runners.

The farmers felt that the scheme flouted the importance of productive enterprises north of the Soutpansberg. Indeed, they believed that Farm Labour Scheme efforts to 'comb the district for any incoming boys' violated a 'gentleman's agreement' that 'the Messina Mine and the farmers at the back [north] of the [Soutpansberg] mountain had first call on

---

[37] Now Tshipise – TshiVenda: 'hot spring'.
[38] TAB HKN 1/1/52 17N3/20/2, 'Foreign Farm Labour Scheme Louis Trichardt Main File.' Department of Native Affairs to Chief Native Commissioner for Northern Areas, Pietersburg, 21 March 1947.
[39] TAB HKN 1/1/52 17N3/20/2, 'Foreign Farm Labour Scheme Louis Trichardt Main File.' Native Commissioner to Chief Native Commissioner, 20 April 1948.
[40] TAB HKN 1/1/52 17N3/20/2 'Foreign Farm Labour Scheme Louis Trichardt Main File.' Secretary of Native Affairs to Chief Native Commissioner, Northern Areas, 26 March 1946.

incoming prohibited immigrants'. The farmers were now a thousand workers short: the aggressive efforts of government recruiters for the depots had caused Rhodesians heading southwards to avoid the area. These workers were needed 'to cultivate the out of season fruit, vegetables, and produce crops, grown by the Njelele farmers, that supplies [sic] the markets throughout the Union, during the "out of season" month, when such crops are not available in any other part'.[41] The Njelele farmers saw themselves as the developers of a distinct agricultural area, making an irreplaceable contribution to South Africa because of their location in the north, and with a labour hinterland that needed protecting. The Farm Labour Scheme appeared to treat the Limpopo Valley, once again, simply as a zone for border policing and for supplying labour to other regions of the country. Responses to the Njelele farmers in 1948 confirmed this. The Native Commissioner at Louis Trichardt claimed that he was 'unaware of any agreement whereby the Messina Mine and the farmers at the back of the mountain were to be given first call on incoming Native Prohibited Migrants'.

Both farmers and state officials presented the arrangements they proposed as consonant with migrants' choices, each side thereby revealing its perspective on the area. Njelele farmers argued that, rather than be diverted by government employees, 'prohibited immigrants... should be allowed to come in the Zoutpansberg and to work freely'.[42] In this view, workers were heading for the farms – local centres of employment – when they were diverted. The Native Commissioner similarly framed this case as an issue of the freedom of labour, but with rather different conclusions. Noting the large amount of land under irrigation, the government dam, and the good climate, the Native Commissioner alleged that the real problem was low wages. 'They are... in a position to afford relatively high wages for their Native labour. Instead they wish to maintain a low wage level at the expense of other farming communities.'[43] By diverting unknowing recruits to their underpaying enterprises, Njelele farmers denied other employers the workforces they deserved as better remunerators, and used workers' ignorance about proper wages to upset the free market of labour. The centres at which migrants chose to work, given sufficient information, were not at Njelele; they were avoiding the farms of their own accord. These points were endorsed by the Chief Native

---

[41] TAB HKN 1/1/52 17N3/20/2, 'Foreign Farm Labour Scheme Louis Trichardt Main File.' Charles Chamberlain, Soutpansberg Farmers' Union to Native Commissioner, Louis Trichardt, 25 February 1948.
[42] Ibid.
[43] TAB HKN 1/1/52 17N3/20/2, 'Foreign Farm Labour Scheme Louis Trichardt Main File.' Native Commissioner, Louis Trichardt, to Chief Native Commissioner, Pietersburg, 25 February 1948.

Commissioner and the Secretary for Native Affairs.[44] Indeed, the Native Commissioner contested the farmers' portrayal of the Farm Labour Scheme. 'It is untrue that the District is combed for Extra-Union Natives by the organisation,' he wrote to the Chief Native Commissioner. 'Extra-Union Natives entering this district at the Border who are contacted there by the Native employees at the subsidiary depots have the said scheme explained to them and are left free to choose whether they will accept employment thereunder or find employment for themselves.'[45] The Farm Labour Scheme itself was part of the range of choices available to migrants.

This disagreement offers a window onto Limpopo Valley agriculture and state officials' views of migration and labour supply in the late 1940s. But, to make sense of Njelele farmers' assertions of their own area as a centre of production with 'first call' on migrants from their hinterland, we have to appreciate what the Farm Labour Scheme was. It was, as Bradford (1993) argues, the latest state attempt to engage in labour recruitment for other parts of South Africa, notably the eastern Transvaal. Such recruitment was also, I contend, an attempt by state institutions to assert territorial control of the northern frontier, in a manner that cast the Limpopo Valley as a border zone rather than a hub of capitalist production. The object was 'to bring about as far as possible the effective rounding up of foreign Natives entering into the Union across the Rhodesian and the Bechuanaland borders'.[46] I first briefly explain the 1940s Farm Labour Scheme, before turning to the wider trend in government border policy from which it emerged.

While it was true that the labour depots did not 'comb the district' for migrants, this was less because of a benign model of worker choice than because of an enduring lack of government capacity in the area. Contrary to the Native Commissioner's description of the Farm Labour Scheme, above, to be brought to a depot was, in effect, to be arrested. Instructions on running the Louis Trichardt depot for the local Native Commissioner stipulated that

Immediately any foreign Native is received into the main depot he must be served with a prohibition notice.... Immediately after service of prohibition notices on a batch of Natives, they should be called upon to elect whether they desire to enter into contracts of employment in terms of the scheme or whether they desire

---

[44] Ibid; TAB HKN 1/1/52 17N3/20/2, 'Foreign Farm Labour Scheme Louis Trichardt Main File.' Secretary for Native Affairs to Chief Native Commissioner, Pietersburg, 9 April 1948.
[45] TAB HKN 1/1/52 17N3/20/2, 'Foreign Farm Labour Scheme Louis Trichardt Main File.' Native Commissioner, Louis Trichardt, to Chief Native Commissioner, Pietersburg, 25 February 1948.
[46] TAB HKN 1/1/52 17N3/20/2, 'Depots: Foreign Labour Recruiting Scheme.' Secretary of Native Affairs to Chief Native Commissioner, Pietersburg, 10 July 1947.

to be repatriated. Any Native who has refused to accept employment should be transported to the border and ordered to return to his country of origin.[47]

Despite considerable efforts, the depot system was a failure. Foreigners from the north went to great lengths to avoid the coercive scheme, which managed to deliver a mere 1,200 workers to farmers during its operation (Bradford 1993: 112). And big farmers elsewhere in South Africa responded to the scheme's lack of effectiveness by turning back to means of recruiting northerners on the border that recalled earlier banditry. Migrants were 'not merely being assaulted, or robbed, or subjected to touts impersonating the police. Some were also being recruited at gunpoint, hunted like ostriches, or captured by dogs before being stuffed into lorries' (ibid.: 115). In recognition of these failings, the scheme was closed in 1949.

Nevertheless, the logic underpinning it continued. By 1953, as apartheid under the National Party approached its fifth year, an infrastructure of depots became part of the wider system to control black movement, while the prison system became a primary source of state-provided agricultural labour. When the establishment of a new 'reception depot' at Nigel, east of Johannesburg, was proposed, it was to be 'a place of detention under the provisions of the Immigration Act and any foreign Native can therefore be detained therein in accordance with the said act'. This new depot was to house black non-South Africans who were unemployed and lacked documents for the Union of South Africa or for urban areas. Such 'recruitment' was a means to police illegal immigration while providing labour to farmers. As the Secretary of Native Affairs put it, 'It is not the intention to repatriate these Natives, but the Department is anxious to gain effective control over their movements and to offer them such employment in rural areas as may be available.'[48]

The two phrases used by the Secretary of Native Affairs in 1953 – 'gain effective control over their movements' and 'offer them ... employment in rural areas' – are both key to understanding the issues at stake in migration across the Limpopo, as these were seen by mid-century South African officials. Indeed, they reflect the dual imperatives that had motivated such policies for decades. In the 1930s, the Immigration Department had developed a special scheme for undocumented migrants, in which anyone caught was given a temporary permit and then made to choose between a year's paid farm work, with deductions for ultimate

---

[47] TAB HKN 1/1/52 17N3/20/2, 'Foreign Farm Labour Scheme Louis Trichardt Main File.' 'Farm labour scheme: special instructions for the Native Commissioner, Louis Trichardt.' Secretary of Native Affairs to Chief Native Commissioner, Northern Areas, 3 March 1947.
[48] TAB HKN 1/1/52 17N3/20/2, 'Foreign Labour Recruiting Schemes – Louis Trichardt.' Secretary for Native Affairs to Commissioner, South African Police, 18 April 1953.

deportation; 'immediate repatriation at one's own expense'; and 'immediate imprisonment and unpaid prison labour' (MacDonald 2012: 283). But the depots discussed in this section bore yet greater resemblance to the efforts of the 1920s. The next section therefore takes a longer perspective on these trends, preceding the rise of Njelele farming.

Policemen or Recruiters? State Intervention on the Northern Frontier

In 1921, the number of Rhodesians heading southwards was rising. The Acting Director of Native Labour wrote to the Secretary for Native Affairs that 'we are somewhat embarrassed with large numbers of these [tropical] natives coming through for employment'.[49] He noted, on the other hand, their 'considerable value' to maize farmers on the Highveld. He recommended that employers be required to provide 'two blankets, a jersey and pair of trousers or other equivalent clothing for every native engaged between April and August inclusive'. So, on the one hand, non-South Africans were entering the country in large numbers;[50] but, on the other, this could be turned into the basis of a system of labour supply with a veneer of state direction.

Neither spikes in numbers of Rhodesian migrants nor their diversion onto white farms were new. Messina Mine had been struggling to secure adequate labour at times in the previous decade, as we have seen, but a lot of Rhodesians had been crossing the border. They were a source of vulnerable, easily exploitable labour, for anything from Natal collieries[51] to public roads.[52] In April 1917, a representative of the Secretary of Native Affairs sent a telegram to the Rhodesian Chief Native Commissioner in Salisbury, reporting that, 'owing to pressure brought to bear on [the natives] by Rhodesian Government officials to repay grain advances made last year large numbers of tropical Natives from Rhodesia have entered Northern Transvaal in search of work'. He suggested that 'there is serious shortage of farm labour in Zoutpansberg and other districts [in the] Transvaal [and] these Natives might usefully be employed on farms'.[53]

---

[49] SAB NTS 2025 26/280 2598/14/473. Acting Director of Native Labour to Secretary for Native Affairs, 18 August 1921.

[50] By 1922, among the 'rush of natives to Johannesburg in search of work' there were also large numbers of people from the 'outside Districts of the Union', notably Sibasa in the far north, as severe economic conditions affected rural areas. SAB NTS 2025 26/280 2598/14/473. Chief Pass Officer, Johannesburg, to Director of Native Labour, 2 September 1922.

[51] SAB NTS 2025 26/280 2598/14/473. Vryheid Railway Coal & Iron to Director of Native Labour, 14 September 1915; Acting Director of Native Labour to Secretary for Native Affairs, 23 September 1915; Secretary for Native Affairs to Rhodesian Chief Native Commissioner, Salisbury, 6 January 1916.

[52] SAB NTS 2025 26/280 2598/14/473. Magistrate, Nylstroom, to Superintendent Roads and Local Works, Pretoria, 3 August 1920.

[53] SAB NTS 2025 26/280 2598/14/473. 'Natives.' O.H.M.S. for Secretary for Native Affairs to Chief Native Commissioner, Salisbury, 11 April 1917.

State attempts to control the movement of northerners by gathering them and supplying them as labour to farms require some contextualisation. The recipients of this labour were on the whole the *groot boere* ('big farmers') of the Eastern Transvaal. These 'progressive' farmers, predominantly of Russian Jewish descent, had vast areas planted with maize, and turned large profits. Unlike the smaller Afrikaner farmers around them, with their black tenants, these were true agrarian capitalists, seeking out the cheapest possible wage labour. As in many capitalist operations worldwide, however, this labour was not 'free' but consisted of foreign migrants acquired under coercive conditions (Bradford 1993). Time and again, state officials would strive to provide them with this cheap labour, drawing on foreign migrants, prison labour, and children. And, time and again, state officials' plans were thwarted by two things. One was the activity of private – often illicit – recruiters, and the other was the resistance of migrants to state intervention (ibid.).

Throughout the twentieth century, many black South Africans remained averse to wage work on farms, preferring the better paid and more masculine option of mining and industrial employment (James 2001). Another reason they left the farms, paralleling the experience of Messina Mine, was poor workers' conditions (as, indeed, the Native Commissioner at Louis Trichardt would allege of the Njelele farms in the 1940s).[54]

Unlike the gold mines, farmers could not make use of the highly institutionalised infrastructure of the Witwatersrand Native Labour Association to recruit and transport workers. Even Messina Mine, which struggled to make up its complement of employees in its first 20 years of operation, was by the early 1930s drawing in workers from much further afield. Messina's accident books from 1931 to 1934 show that most of workers hailed from areas well to the north of the border, especially from Nyasaland and Northern Rhodesia.[55] Many farmers, though facing the same acute constraints on the availability of labour, remained dependent on more or less informal means of labour recruitment.

By 1922, the Director of Native Labour was pushing to regularise arrangements, making use of existing labour recruiters.[56] A Government Notice in 1918 had prohibited agents from recruiting Rhodesians in the Union.[57] Meanwhile, a conference in Pietersburg the same year, with Southern Rhodesia's Chief Native Commissioner, had 'arranged that

---

[54] SAB NTS 2025 26/280 2598/14/473. 'Native Labour Supply.'
[55] My thanks to Jack Klaff, who acquired the books when the mine closed in the early 1990s and who made them available to me for my research.
[56] SAB NTS 2025 26/280 2598/14/473. Director of Native Labour to Secretary for Native Affairs, 13 December 1922.
[57] Section 10 of Government Notice No. 1693 (1918). This refers to active recruitment, as opposed to employment, and was a response to illicit labour agents' activities on the border.

short of repatriation the recruitment of boys who filtered through would not be allowed but that as far as possible we would place them on farms and industries in the Northern Transvaal'.[58] Prefiguring later schemes, migrants would be brought to a depot at Louis Trichardt and allocated to prospective employers. But such limits were entirely unrealistic. In any case, Rhodesians were in South Africa seeking work well beyond the Northern Transvaal. As the Director of Native Labour noted, 'it is not possible to repatriate them and they cannot all be absorbed here [Johannesburg], and unless an outlet is found for them they will inevitably drift to slum areas and become criminals.[59] Recruitment on behalf of farmers was therefore provisionally condoned on the Rand, too.[60]

The result was an uneasy compromise, which became the enduring pattern in state attempts to control labour procurement. On the one hand, recruitment created a labour demand, inviting an increased supply of migrants; on the other, the government hoped – at least officially – to discourage migration.[61] Tensions in the arrangements were reflected in the regularity with which the Secretary for Native Affairs emphasised to farmers applying for labour that 'the policy of the Government is to discourage the immigration of these Natives by all possible means'.[62] He meanwhile referred requests to the Director of Native Labour, who by now was being tasked with allocating Rhodesian workers to prospective employers. The larger farmers discussed by Bradford (1993) themselves had come to rely heavily on the fact of a regular, cheap labour supply from Rhodesia.[63]

The instability of this official position was especially evident when it came to determining who was allowed to do the recruiting. Farmers' own representatives for recruiting Rhodesians did not hold regular recruiter licences, whereas licensed agents were not allowed to sign up Rhodesians because of the 1918 ban. As the Sub-Native Commissioner at Pietersburg explained to the Secretary for Native Affairs, after an encounter with a particularly angry licensed labour agent: 'there is not the slightest doubt that these so-called Agents [those of the farmers] are using Native touts, and the licensed Labour Agent has a just grievance that he, the Labour Agent, although his licence is endorsed, to recruit for various farmers, he

---

[58] SAB NTS 2025 26/280 2598/14/473. 'Native Labour Supply.'
[59] SAB NTS 2025 26/280 2598/14/473. Director of Native Labour to Secretary for Native Affairs, 19 January 1923.
[60] SAB NTS 2025 26/280 2598/14/473. Director of Native Labour to Secretary for Native Affairs, 1 February 1923.
[61] SAB NTS 2025 26/280 2598/14/473. Secretary for Native Affairs to W. H. Rood, 2 October 1924.
[62] E.g. SAB NTS 2025 26/280 2598/14/473. Secretary for Native Affairs to Lowveld Farmers Association, Nelspruit, 3 October 1924.
[63] SAB NTS 2025 26/280 2598/14/473. Director of Native Labour to Native Sub-Commissioner, Louis Trichardt, 29 October 1924.

is prohibited from handling Rhodesian Natives'.[64] The selective relaxation of rules to meet the needs of labour supply and demand had created the conditions for new 'labour pirates'. As it would repeatedly, the image of the 'bandit' once again reared its head.

Supplying Rhodesians to employers was justified on the basis that governmental institutions were unable to police the border. Yet the alternative to preventing Rhodesians from entering South Africa had come to look no better: using the same kinds of unscrupulous labour agents with whom officials had long been struggling. As one farmers' association put it, 'The Mines have their own Native Labour organisations but the farmers are in the hands of agents who exploit them in a shameful manner.' In this case, recruiters were using farms as pools from which to recruit labour for elsewhere, simply moving workers around for profit. The farmers' association's concerns are reminiscent of those of Messina Mine in the previous decade. The proposed solution was that 'if the demand could be met by a Government organisation controlling the indenture of natives on the lines of the present recruiting organisations, there would be a uniformity of methods, especially if stern measures are legalised to carry out such organisations'.[65]

Indeed, a model built on government labour depots – whose 1940s version was discussed in the previous section – was developed precisely as a solution to these problems. The Director of Native Labour was already responsible for allocating Rhodesians to employers through an existing depot at Louis Trichardt managed by a contractor, but in 1925 the Sub-Native Commissioner at Louis Trichardt took over running it directly. Regulations gave the state, in theory, a monopoly in recruiting migrants from the north, while 'clandestinely, with no legal basis, the almost moribund labour bureau at Louis Trichardt was resuscitated and supplied with a semi-private bank account' (Bradford 1993: 103). Indeed, the number of migrants passing through the Louis Trichardt centre – 2,780 in 1925 – meant that capitation costs could be lowered. Officials appeared to be behaving like labour agents off the books, their activities 'approximat[ing] to recruiting' and 'the Government... making a profit out of its dealings in this class of labour'.[66] The Control and Audit Office worried that it looked all too much like a profit-motivated labour recruiter:

Its basis is the improper entry into the Union of prohibited Rhodesian natives, it serves as an employment bureau for farmers and others who are willing to

---

[64] SAB NTS 2025 26/280 2598/14/473. Sub-Native Commissioner, Pietersburg, to Secretary for Native Affairs, 14 July 1923.
[65] SAB NTS 2025 26/280 2598/14/473. Middelburg Distrik Boere Unie to Minister of Labour, 21 November 1924.
[66] SAB NTS 2024 26/280 'Note on Entry of Rhodesian Natives to Union.'

use such natives, and the larger the number of natives apparently the better the results financially it is able to show. Moreover in certain quarters it might be held to encourage the improper entry of such natives because of the ready means it affords of finding employment.[67]

Success was illusory. Numbers dropped off in 1926, and it was suspected that migrants were avoiding the depot by entering the country across a different border.[68] If 1925 was the zenith of the labour depot's operation, by 1929 it was so little used as to be practically irrelevant (Bradford 1993: 105). Once again, limited state capacity to govern territory constrained attempts to intervene in foreign labour recruitment. Like the attempt to establish a labour depot system around Louis Trichardt two decades later, the scheme was a failure. Seen in terms of the nexus of controlling territory and labour, depots were to be hubs – points of coordination and close supervision, where management could be effective. Unsurprisingly, however, it seemed likely that proportionately fewer Rhodesians were being successfully drawn into the depot system, than were avoiding it. A depot could be managed, but also evaded. And it purported to draw labour distribution out of illicit markets, but appeared to participate in those very same markets. Louis Trichardt allocated labour on 12-month contracts to farmers by ballot in the mid-1920s. There was a fine line between administering territory and controlling human mobility, on the one hand, and engaging in profitable labour recruitment on the other.

This case study focuses on the period between the 1910s and around 1950. It does so because in this period Rhodesians became crucial to the growth of 'progressive' agriculture in the Transvaal. With a severe lack of workers, agrarian capitalists came to rely on migrants from Rhodesia for labour. They successfully petitioned state officials to keep them supplied with workers, whether aliens, convicts, or children. The period also shows attempts by senior figures in state institutions to juggle these labour questions with assertions of territorial control. Turning 'clandestine immigrants' into workforces appeared to achieve this, but the same lack of state capacity that made the border ungovernable in the first place made interventions into labour dynamics limited and often easily avoided by migrants and private recruiters.

Both of these efforts – to supply labour to the eastern Transvaal, and to maintain a veneer of territorial control in the unruly far north – sidelined the priorities of capitalists in the Limpopo Valley itself. It was the state's role as a labour recruiter that brought it into conflict with the growing commercial agricultural sector at Njelele. For farmers north of the Soutpansberg, this was not just a zone in which to control movement

---

[67] SAB NTS 2024 26/280 C. A. & G. to Secretary of Native Affairs, 17 February 1926.
[68] SAB NTS 2024 26/280 'Note on Entry of Rhodesian Natives to Union.'

Behind the Mountain 99

by sending Rhodesians to work in the eastern Transvaal. This was a place of its own, a hub of production.

The importance of the Limpopo Valley as labour supplier decreased markedly under apartheid, due not to increased territorial control on the border, but to the provision of local alternatives. These were enabled through the general tightening of controls within South Africa and by schemes providing farmers with unpaid workers from the prison and internal policing system. Black South Africans in the 1950s, hoping for mining contracts, were tricked and coerced into becoming compounded farm labour in appalling conditions, as famously laid bare by *Drum Magazine* exposés (Sampson 2005). The shift from foreign labour was far from complete, of course, and the farms and mines reserved the right to depend on non-South African workers.[69] However, the northern border's role in supplying such labour appears to have changed as a result of tighter territorial control in the core areas of South Africa, beginning with the domestically oriented 'Farm Labour Scheme: Union' in 1947 (Bradford 1993).

The border remained porous, and a flow of labour across it remained crucial for capitalist enterprises in the Limpopo Valley. Detailed research on 1950s and 1960s border control along the Limpopo River is yet to be conducted. But the later use of farmers as a buffer evidences the border's continued permeability. Indeed, older black workers who had been on South Africa's emerging border farms in the 1970s remember significant local mobility. People moved back and forth across the river between the South African agricultural estates and the Rhodesian commercial farms and Tribal Trust Lands (black reserves). In the 1970s and 1980s, Zimbabwean migration to South Africa was related to the marginalisation by the Rhodesian/Zimbabwean state of its southern border areas, and to persecution for alleged dissidence (see introduction to this book; also Mate 2005; Werbner 1991). Whatever its causes, it was on this continued cross-border mobility that the border farmers relied. The final section of the chapter turns to examine how these border farmers have managed, since the 1980s, to achieve a provisional resolution between conflicting perspectives on the area: porous borderland versus established hub of production.

'Special Employment' in the Border Zone

The border farmers, as noted earlier, embodied the central tension in border capitalism: they were intended both to be a buffer in a peripheral part of South Africa, and to develop the area as a hub of work and

---

[69] This is illustrated by the agricultural and mining sectors' outrage at the 1960s Froneman Commission recommendation that they reduce their dependence on non-South African labour. Their protests led the apartheid government to disregard the Commission's findings (Crush 2000: 2).

residence in its own right. Farmers like Jan continue to complain about the impact of border policing on his farming.

However, the farmers along the Limpopo were quickly accorded a special status as border employers, and favoured with arrangements that transcended the centre/periphery divide. In the past, labour migration to South Africa was governed by bilateral treaties with surrounding countries, but there was none established with Zimbabwe. The arrangements put into place for border farmers were a pragmatic response to this lack (Crush 2000: 7). 'From the early 1980s, farmers north of the Soutpansberg recruited Zimbabwean workers under the special exemption provision in the Aliens Control Act' (HRW 2006a: 14). By the 1990s, this arrangement became a distinct 'special employment zone', in which farmers did not have to go through the usual bureaucratic channels to recruit workers (Rutherford 2010: 68). Although the special employment zone was discontinued in the late 1990s, its legacy continues to shape border farming. There is a sense in which the area remains recognised both as 'border zone' and 'special employment zone'. The arrangement represented farmers' success in carving out a border area that was also a hub of capitalist enterprise.

In the special employment zone, which ran from the border to the Soutpansberg, Zimbabweans could enter South Africa legally, so long as they carried a special permit (a 'BI-17') linking them with a specific farmer. This produced a distinct border agricultural economy, in which farmers were able to set their own terms, reinforced by the area's remoteness and norms of white agriculture:

Commercial agriculture within the 'special employment zone' resembles a colonial plantation enclave: (a) it produces for distant markets; (b) resources are used on a land-extensive and labour-intensive basis; (c) labour is employed, accommodated and controlled on the farm property and, (d) the sector is isolated by considerable distances from urban centres. (Lincoln and Maririke 2000: 43)

The constitution of the border area as a recognisable centre of agricultural activity had a number of on-the-ground effects, all of which reinforced farmers' locally powerful positions and shaped the place of the farms on the border today. Firstly, for Zimbabweans crossing the border, the system of BI-17 permits placed a premium on connections with particular farmers and their black foremen. This highly personalised system of labour regularisation meant that networks developed between the South African farms and villages across the river in Zimbabwe, as people sought avenues for employment. Not only did this result in webs of kinship and other relationships emerging both across the border and between the border farm populations; it also led foremen to become key figures in the border area. Both the webs of relationships and the central role of foremen continue to shape the border today.

Secondly, cross-border recruitment within this system led to the opening of an 'informal' border post at Gate 17, a point on the border fence immediately adjacent to the farms. According to my informants – both farmers and farm workers – this was a somewhat *ad hoc* affair. A few South African officials would set up a mobile station on the southern bank of the Limpopo, specifically to process Zimbabweans heading to border farms under the arrangements of the special employment zone. A published eyewitness account of Gate 17 suggests something similar: a handful of Home Affairs officials and South African soldiers; a military tent as an office, with tea brewed by a woman providing an 'informal service' (possibly a domestic worker from a farm); Zimbabwean officials at a station across the river. White farmers from elsewhere sent pick-up trucks to recruit workers (Lincoln and Maririke 2000: 47–8). The border farmers themselves, however, did not even have to send trucks, since the gate lay at the edge of the farms. Gate 17 explicitly brought together border enforcement and the farms' roles as local hubs – important economic centres that required special access to labour. Today, although the 'informal' border post has disappeared, it has been replaced by similar, though even more irregular, arrangements. Workers often go through the border fence at the garrisons, with soldiers' permission, or simply use the farms' own gates, which lead to water pumps in the river. Border officials accept that recruits come straight through the fence, bypassing conventional forms of state control.

Thirdly, within this system, it was considered common practice for recruits to make their way to the farms without papers (in this case avoiding Gate 17), where they would in theory be issued with a permit (Lincoln and Maririke 2000). The risk of arrest and deportation when undocumented was borne entirely by the workers themselves. Some farmers, indeed, would not document seasonal workers at all. Migrants crossed clandestinely for several reasons. One was that doing so was fairly straightforward, because border enforcement remained patchy. Another was that what risk there was of deportation was diminished once migrants reached the farms. For the estates were considered 'protected ground', on which one would not be arrested. A third was that, on arrival at the farms, *post hoc* legalisation was easy, and indeed remained the most convenient method of documentation. Moreover, many migrants, as in the past, did not yet know their ultimate destinations (Crush 2000: 9). Farmers welcomed this state of affairs, because it meant recruits would turn up ready to work on the first day of the harvest with a minimum of fuss. Documentation, on the other hand, remained somewhat cumbersome. Further, this 'grey' employment status made for a vulnerable, easily exploitable workforce (ibid.: 10). Today, precisely these expectations about employment, documentation, and risk shape harvest-time recruits' lives on the farms, as they sleep in the bush for fear of police raids.

The special employment zone was discontinued in the late 1990s because of the post-apartheid media and NGO outcry 'that "national sovereignty" be enforced, particularly against what they viewed as an egregious continuation of the authority of white (Afrikaner) farmers in post-Apartheid South Africa' (Rutherford 2010: 69). However, these wider anti-farmer sentiments had minimal impact on the border farms, and the notion that farmers had a central place in controlling local territory and mobility persisted. There was a brief hiatus in the state-sanctioned recruitment of Zimbabweans, but this itself revealed farmers' considerable autonomy. In 1999, farmers in the area, who employed around 15,000 Zimbabweans, were ordered by the Department of Home Affairs to recruit South Africans instead (HRW 2006a: 14). Farmers I spoke to reported having established a recruitment bureau at Messina, but claimed that no South Africans came forward. They cited their own remoteness and black South Africans' aversion to agricultural labour as reasons for this. On these grounds, they refused to stop employing Zimbabweans. Quickly, ways were found to continue border agriculture's special arrangements by other means.

The Transvaal Agricultural Union brokered an 'informal agreement' with the Department of Home Affairs in 2001, enabling farmers north of the Soutpansberg to employ Zimbabweans according to the same labour rights as South African workers. In 2004, a Memorandum of Understanding between Zimbabwe and South Africa extended this to the whole of Limpopo Province. Under the Memorandum, Zimbabwean farm workers were issued with Emergency Travel Documents, in contrast to other Zimbabwean migrants who had to apply for passports. This arrangement remained in place during the period of fieldwork, at the height of the Zimbabwean crisis.

From 2005, and throughout the period of fieldwork, border farm recruitment was brought into line with standard South African immigration law. Farmers apply for corporate permits, under which a fixed number of foreign workers can be recruited. Corporate permits are issued after employers demonstrate that they are unable to find South Africans to do the work. In practice, however, this leaves ample room for border farmers to be accorded special privileges because of their location: 'proof of the need to employ the requested number of foreigners [is] ... evidently a formality' (HRW 2006a: 14). In a more diffuse sense, local authorities continue to acknowledge farmers' centrality in the border's rural areas. Established notions of farmer 'sovereignty' over their land and labour combine with established local arrangements to accord white landowners considerable autonomy from, and freedom to ignore, border authorities (Rutherford and Addison 2007; Rutherford 2008).

Returning to the complaints of Jan and other farmers mentioned at the beginning of this chapter, their predicament must be understood in terms of the widespread recognition of farmers' place on the border. It is true that white farmers along the border today fear police reprisals for harbouring 'illegals'. They feel that they attract little affection from the now-black police force. They are nevertheless in a position to meet with senior police and army officials to negotiate the status of their workers. One white, high-ranking soldier regularly stops by the border estates to speak with farmers, with whom he is on good terms. And I even heard widespread talk among workers of a neighbouring farmer who had simply blocked the gate to his land and refused the police entry. Farms make laminated employment cards, decorated with Clipart pictures of fruit trees, and issue them to workers. Police and soldiers generally accept these from workers on the farms themselves, although often not when they are caught off the estates. *Ad hoc* agreements between farmers, police, and army, while only partially and locally effective, mean that farm employment lends some degree of predictability and safety to everyday lives. The provisional security afforded by proof of connection to the farms is so valuable for diverse livelihood strategies that a market in forged employment cards developed in the Zimbabwean border town of Beitbridge. In the century-old conflict between views of the border area as hub or periphery, border farmers have managed to assert their own position, with the farms constantly shaping border policing, and not merely the other way around.

Conclusion

Different instances of white enterprise in the Limpopo Valley, occurring at various points over the last century, give some insight into the enduring motivations and problems of such enterprises, each with its own relationships to the South African state's attempts to police the area as a border zone. In each case, it was crucial to establish a sense of the Limpopo Valley as a place in its own right, countering the state's attempt to view the area as little more than a frontier for border policing. This was a struggle over the meaning of territory. Such assertions by border capitalists were struggles over local control of labour recruitment, government, and simply everyday life.

Understanding these struggles puts today's border farmers' experiences into perspective. They contend with border policing, to be sure. Sometimes, their undocumented workers are deported, hindering production. However, what in fact exists along the border is a compromise that favours farmers. Through this compromise, the farms' local roles are both recognised and tacitly supported.

Centres of production have profoundly shaped the region. Although it closed in the 1990s, Messina Mine was the basis for what is now Musina town, which is vital to cross-border trade and to the border's farms

104   Zimbabwe's Migrants and South Africa's Border Farms

(it has also become the headquarters of a nearby De Beers diamond mine). The farms, in turn, have become social centres of gravity along South Africa's northern border. Exploring the farms' roles in farm workers' lives is the subject of the remainder of this book. To that end, the next chapter focuses on the farm of Grootplaas. It shows how black workers regard the farm not only as site of employment, but also as a place in which their lives are anchored by homes and relationships.

4 Producing Permanence
Employment and Domesticity in the Black Workforce

Introduction

The border farmers achieve an ongoing compromise with soldiers and police, in which they are recognised as key figures managing territory and labour on South Africa's northern boundary. Their autonomy as farmer-landowners reflects colonial- and apartheid-era norms (see du Toit 1993). As the previous chapter noted, the 'special employment zone' of the 1990s made this association with the past too explicit, and consequently drew opprobrium. Farmer sovereignty has been perpetuated, however, through informal, on-the-ground understandings with border authorities and Home Affairs officials. Thus undocumented workers awaiting permits are often left alone by deportation teams. The arrangements represent one way to resolve the age-old conundrum of capitalists in the area: how to recruit and control workforces in a place that is also an aggressively policed borderland.

Whereas the previous chapter described the arrangements secured by white farmers, this chapter turns to ways that black farm workers themselves organise their lives. Recent anthropological studies of farms in the region north of the Soutpansberg (the border itself and Tshipise to the southeast) see the circumstances just described as leaving farm workers acutely vulnerable (Rutherford and Addison 2007; Rutherford 2008, 2010). Farmers' agreements with the authorities trap workers between overlapping forms of territorial control, rendering them highly exploitable. Border police deport workers. White farmers, attempting to increase 'efficiency' in an environment of competitive neo-liberalism, use their sovereignty over land and workers to secure the cheapest and most docile labour available. Farmers regularly underpay, or dismiss workers without due process. 'Undocumented migrants are attractive to farmers because they are easily accessible and disposable virtually on demand,' claims one study of the area, 'and are vulnerable to a wide range of abuse and exploitation' (Rutherford and Addison 2007: 625; see also HRW 2006a).[1]

---

[1] On one border farm to the east, close to the Kruger National Park, workers appear to have such insecure prospects that their lives become dominated by acquisitive, atomised hustling. Few black residents invest in farm life, and their single-minded projects to

It is certainly true that seasonal recruits on the farms are extremely vulnerable. Until agreements are worked out between farmers, police, and the military each year, the compounds are perfect targets for deportation teams, full of Zimbabweans whose permits are delayed by bureaucratic inefficiency. At the beginning of the harvest, many workers regularly sleep in the bush, afraid that they will be apprehended if they live in compound housing. Others are locked inside their accommodation by friends, while padlocks on the outside of their doors make their rooms appear empty. In my experience, farmers are generally inconvenienced by the disappearance of their workers after deportation raids. One reason is that recruiting replacements brings added paperwork. South African working visas must be affixed to Zimbabweans' identity documents.[2] In any case, deported workers generally return to the farms within a couple of days. Having crossed back again through the border fence after being dropped at Beitbridge, they return to the farm to demand their jobs and wages. However, farmers also clearly benefit from workers' vulnerability. Seasonal workers' fear of police on the roads keeps them confined to the labour compounds. In local horror stories, farmers have even been known to manipulate the boundaries between their own territorial authority and that of the police. After a month of work, during which police stay away, farmers contact the authorities, reporting undocumented Zimbabweans on their farms; they thereby avoid paying for labour. Similar stories are told of farmers on the border between Mozambique and South Africa.

However, there is another side to life on the border farms: one of sociability, models of ethical responsibility, and permanence. At Grootplaas, core workers spend considerable time and resources developing their domestic arrangements, including their accommodation. The working populations built around them are bound together through kinship, friendship, and sexual relationships. Such connections also bind together the whole of this border farming community, including the one remaining commercial farm on the Zimbabwean side. Although the border farms lie on an important migration route, they are remote and surrounded by game farms. Day-to-day existence, for most, generally revolves around the string of labour compounds, between which residents travel by bicycle or on foot. Compound residents along the border are also drawn together by their interest in the latest Zimbabwean news. Every evening, radios across the compounds whine with the siren-sounds of the Zimbabwean government's scramblers, as people tune into opposition radio stations.

---

build homesteads in Zimbabwe make for an individuated workforce. This state of affairs is underpinned by the fact that all black employees on that farm have seasonal contracts (see Addison 2013).

[2] As discussed in the previous chapter, these were ETDs (Emergency Travel Documents) at the time of fieldwork – substitutes for Zimbabwean passports that were developed to streamline the documentation of farm workers in South Africa. More recently, migrants' access to Zimbabwean passports has improved.

Visits to town are followed by exchanges of newspapers. Work-time conversations debate the latest events. Many residents are sharply critical of the Mugabe regime, but discussion, especially among permanent workers, is often focused on political intrigue rather than the horrors of state brutality – a sign of workers' distance from the Zimbabwean setting.

The farms act as local hubs, in an area where the border is porous and the compounds represent unusually large, waged populations. These communities consequently draw in many people from the surrounding region. Payday markets attract South African and cross-border trade. Residents of Musina have relatives and acquaintances on the farms. One senior permanent worker at Grootplaas is a bishop in the United African Apostolic Church, known to members of the town congregation. During the harvest, when farm worker populations are at their peak, local music stars play gigs in improvised arenas with a cover charge levied at the entrance, all organised collaboratively by a Grootplaas permanent worker and a border customs clearance agent from Musina who has a girlfriend in the compound.

These kinds of interaction rely on the cooperation of border guards. Thus understandings exist not only between farmers and state officials, but also more directly and locally between workers and the soldiers who man the garrisons. The farm area's remoteness and its tight communities localise the supposedly impersonal military presence of the border. The army patrols are far from their regiments during their three-month border stints. Soldiers, housed in groups of about ten in tiny buildings, complain of loneliness and boredom. Some are sharply aware that preventing border jumping is a futile exercise: the fence is long and permeable, and those caught and deported only try to cross again. Many soldiers go to nearby farm compounds for company and sex. They often wander through the Grootplaas compound, looking for women to sleep with in exchange for their tinned rations. Some establish longer-term continuity in their relationships with women. At one point, attempting to win favour with the border population, the soldiers even fielded a football team in an inter-farm tournament.

Both the more widely held stereotype of working life in this region – of a vulnerable, fugitive existence – and the second – that the farms foster sociability and permanence – are true. What the view from Grootplaas reveals is a sharp difference in this regard between the lives of established permanent-worker residents – *mapermanent* – and their seasonal counterparts. This chapter adds a level to this study of border farming by demonstrating the contrasting experiences of workers according to their place in a farm workforce. It asks: what distinguishes the lives of *mapermanent* from those of the less-established black residents of Grootplaas?

At one level, the differences between permanent and seasonal workers signal the effects of casualisation, in which the majority of workers become increasingly insecure, with the effect of serving the 'needs of capital'. The casualisation of farm work in South Africa is one consequence of post-apartheid changes in agriculture, what Rutherford and Addison call a 'fundamental class project associated with the current phase of neo-liberal restructuring' (2007: 626). The effect of these changes has been to draw a strong distinction between small, secure core workforces and floating labour reserves with few rights (Ewert and du Toit 2005).

However, this chapter goes beyond such accounts. Distinctions between permanent and seasonal workers are not simply the transparent results of 'neo-liberalism'. Inspired by classic situational analyses, a focus on micro-dynamics in a resident workforce reveals how workers themselves create distinctions. Different positions in the workforce become resources that workers use to shape diverse aspects of life on the farm. This is similar to Rutherford's (2001) ethnography of Zimbabwean farm workers, which also analyses workers' dependencies in relation to their work roles. In Rutherford's study, senior workers use their positions to control farm residents' access to credit granted by white farmer-employers. Where his case and mine differ, however, is in the degree of fragmentation in the Grootplaas population.

At Grootplaas, *mapermanent* take advantage of their relative security among transient, often vulnerable, migrants. Understanding the work-related temporalities that shape labour regimes – permanence versus transience – takes us beyond the simple point that casualisation relies on core workforces, and beyond employment categories. It takes us into the relationships and forms of emplacement that determine sharply divergent modes of domestic life. As workers, local notables, and established residents, *mapermanent* offer a first layer of insight into how the border farm populations structure a fragmented environment.

The chapter therefore begins with distinctions between permanent and seasonal workers at work itself, before tracing those differences into life in the labour compound outside work time. This reveals that permanent employees' lives are ultimately distinguished by the kinds of relationships with women and soldiers that they are able to forge because of their privileged positions. A range of forms of inequality are organised around the central place of *mapermanent* at Grootplaas. The examination of unequal relations in the compound, presented in this chapter, paves the way for closer attention to the perspectives of different compound residents in the chapters that follow.

Mapermanent and Seasonal Recruits: The Working Context

Grootplaas employs around 140 black permanent employees, mostly Zimbabwean men, to tend orchards, maintain irrigation and water-pump systems, and carry out a range of other tasks throughout the year; women are only permanently employed for domestic work in the farm offices and white houses. The vast majority are in their twenties, thirties, and forties, with a few of the senior members of the workforce in their fifties and older. Among 78 permanent workers surveyed in 2008,[3] the mean age was 37; the youngest worker was 23, and the oldest a veteran security guard of 62. *Mapermanent* work according to irregular, task-based regimes with considerable autonomy, living on farms in some cases for the whole of their working lives.

Unsurprisingly given the border's history of mobility, described in the previous chapter, Grootplaas's population is diverse. Demographic data from interviews conducted in 2008 give some indication. Twenty-four of seventy-nine *mapermanent* described themselves as Venda, of whom three were South African and the remainder Zimbabwean. Fifteen were Ndebele, seven Shangaan, six Sotho, one Pfumbi, one Chewa (a Zimbabwean from Harare), and two South African Northern Sotho. Twenty-three of the sample were Shona, mostly Karanga from Masvingo Province.

However, these numbers disguise as much as they reveal, masking the complexity underlying them. In a mixed region like the border, ethnic distinctions are not sharp. People speak the languages of their neighbours. Many Sotho hail from Venda-dominated areas around Beitbridge and up towards Gwanda, and they grew up speaking TshiVenda as well as SeSotho (the same goes for the one Pfumbi respondent). The very notion of ethnic affiliation must be approached with caution: identification may follow paternal descent, yet first language may follow residence or mother's background. Marula, the South African foreman, is an important example. He had a Shangaan father, but he was raised speaking TshiVenda on the Venda-dominated border farms. The kinship networks that now shape his life connect him to Venda villages across the border. TshiVenda speakers, in turn, have learned other languages while working in Beitbridge or on the farms – or, in the case of learning ChiShona, through interactions with the Zimbabwean state.

Recent economic and political troubles north of the border have led larger numbers of people from all over Zimbabwe to seek work in South Africa, increasing the proportion of ChiShona speakers. The majority of Shona *mapermanent* (13 out of the 23 surveyed) have come to the

---

[3] Interviews were conducted with 105 permanent workers and dependants. Of these, 79 were on fully permanent contracts. Interviews were conducted in the compound, following a questionnaire, and with the help of a research assistant. It was not possible to use random sampling techniques. But the sample analysed here represents more than half of the total population, and so gives at least some sense of the characteristics of the permanent workforce. The sample for age is 78 rather than 79 because one respondent did not supply the necessary information.

farm since the escalation of the Zimbabwean crisis in 2000, whereas the majority of Zimbabwean Venda (13 out of 21) had come before 2000. Nevertheless, it is important not to overstate this difference. There were many Shona permanent workers on the farms before Zimbabwe's political and economic situation deteriorated. A more dramatic shift, in fact, is among Ndebele: only four of the 15 Ndebele workers surveyed came before 2000.

The border farms are extremely multilingual places, but it is TshiVenda that occupies centre-stage. The most senior and the most visible figures in the compound speak it, and it is the language of authority to which others must accommodate themselves. The core of the population are those who grew up in the border area, have a history of work on the farms, and enjoy support from dense cross-border networks. In the farm's earlier days, labour demands were met through the recruitment efforts of particular black workers from villages across the fence. Many workers – permanent and seasonal – still hail from these villages. Reflecting this, 15 of the 24 Venda permanent workers surveyed, and five of the six Sotho, had kin at Grootplaas when they were recruited; only eight of 23 Shona *mapermanent* joined relatives when they came to the farm. People from the border area cross into Zimbabwe regularly to visit. By contrast, some *mapermanent* from further afield in Zimbabwe visit home only very rarely – a result not only of the distance and expense involved in travel, but also of their commitment to established lives at Grootplaas itself. For all *mapermanent*, however, the significant investments they make in their lives at Grootplaas, described below, compete with the necessity to send money and goods to kin in Zimbabwe.

In April and June, Grootplaas employs up to 460 seasonal workers – mostly women in the packshed, and a mostly male workforce of pickers in the orchards. It is among these employees that the effects of the Zimbabwean crisis are most obvious. Given high labour mobility and large-scale Zimbabwean displacement through the area, the farm's workforce is extremely varied in terms of patterns of movement and settlement. Many seasonal workers arrive as strangers on the farm, forming part of a flow of 'unknown people' through the area. For some transient migrants from Zimbabwe, their time as migrant farm labourers is categorically different from their previous lives, and they experience it as exile during a period of exceptional crisis – hyperinflation and political degeneration – at home. People of very different ages are affected: although the majority of seasonal workers are in their twenties and thirties, a conspicuous minority are older, while others are in their teens. In 2007, many, across the age range, were engaging in farm labour for the first time. This sense of upheaval, itself explored in greater depth in the next chapter, is all the more marked because of the ways seasonal workers are sharply distinguished from *mapermanent* in the labour process. Seasonal employees' employment status is manifest in diverse aspects of daily life, all of which

Figure 4.1. *Mapermanent* workers supervising the picking operation.

make for a very different experience to that of the core workforce. But the most visible thing setting *mapermanent* and seasonal workers apart is the nature of their respective work.

Seasonal labour, tightly structured and intensive, comes as something of a shock to the inexperienced. Picking in the orchards is carried out mainly by men, thirty to a team. Supervisors, drawn from the ranks of the *mapermanent*, maintain the work pace by unabated shouting. But overseers also step back from the picking teams, joke among themselves or with particular pickers, and even consult a newspaper for interesting stories. Picking work is analysed in detail in the next chapter. For now, what is important to note is that, by choosing the extent to which they are engaged in the picking process at any particular moment, *mapermanent* underline their difference from seasonal recruits (see Figure 4.1).

A similar distinction prevails in the packshed, a single, enormous room with a high, corrugated-metal roof. In this space – almost bereft of natural light, despite a few windows – are two conveyor systems for washing and grading fruit (the conveyors come together for packing). The hall of the packshed appears as a mass of green-and-yellow machinery, staffed by rows of women at their stations. Before it reaches the workers, the fruit is tipped from trailers into large, shallow baths of warm water for an initial wash, drawn onto conveyor belts by mechanical paddles, and rinsed by automated spraying machines. Beyond this point wait the pre-graders, the least-experienced of the packshed's women, at benches on either side of the conveyors. The conveyors at this stage are partitioned into three tracks, demarcated by metal barriers, and the fruit runs down the two outer tracks two or three at a time. Pre-graders roll the oranges or grapefruit up and down under fluorescent strip-lights, checking them for blemishes or green skins which would make them unsaleable. Blemished fruit is placed in the middle part of the conveyor, between a pre-grader's station and that of her opposite number. From here it is diverted to a huge cage outside the packshed for sale to local juice makers. Green fruit is thrown down a chute located next to each pre-grader, from where it is conveyed to the ripening room and treated with chemicals. Fruit that passes the pre-graders' inspection is coated with preservative chemicals and a thin layer of wax by machine, before passing to other graders who follow a very similar process to the one just described, this time deciding whether each item is Grade A or B depending on the extent of external blemishes. The fruit then moves on to packing, where each packer has a station continually supplied with fruit and cardboard boxes – the boxes are hooked onto a moving chain overhead by other workers in a separate section. Packing is according to size, and Grade A citrus is first run through an optical machine, which photographs each item to calculate its precise volume.

At all these stages of the packshed process, work is monotonous and regular in comparison to picking in the orchards. Paid by the hour, its pace is set by the conveyors and other machines. The machines themselves are so loud that conversing while working is difficult. One grader highlighted falling asleep as the primary challenge to endurance in grading work, in contrast to the physical exhaustion of picking. But while picking is bounded by a fairly well-defined working day – all work begins at seven o'clock in the morning – packshed hours often extend past ten o'clock in the evening to process backlogs of trailers. The graders, at least, have the advantage of being able to sit down; packers have to stand.

In the packshed, as in the orchards, most of the permanently employed minority carry out supervisory or administrative roles. Above the machinery is a system of gantries from where it is possible to observe all work with a considerable degree of precision. At the central point of the gantry system is the black personnel manager's office, where he and other permanent employees do administrative work (see Figure 4.2). The black packshed manager wanders the gantries, keeping an eye on the work process and the machines, while the grading foreman and forewomen[4]

---

[4] These lower-level packshed supervisors – the foreman and forewomen – were not *mapermanent* in 2007, although the foreman was later recruited into the core workforce. Forewomen have well-established connections to Grootplaas and come year after year. One,

Figure 4.2. Grootplaas packshed, showing gantries and personnel manager's office.

monitor the quality of work at closer range, often coming off the walkways to patrol along the grading rows. The farmer himself regularly wanders around the gantries, flouting his own packshed rules by smoking a cigarette. Below, the women on the conveyors wear *doeks* (headscarves) and aprons, both fulfilling hygiene requirements and conforming to a style worn generally by black women working for white employers in Zimbabwe and South Africa. Distinctions of dress and positioning within the packshed are further reflected in the often sharp separation between seasonal and permanent workers when they sit outside to eat their packed lunches.

Seasonal workers labour in closely observed and regulated environments, pushed by piece rates or conveyors to process sufficient quantities of fruit to make buying-agent deadlines. *Mapermanent*, as we have seen, are less bound by these work regimes. As supervisors, they are able to dip in and out of the work processes – picking supervisors, especially, are always ready for a chat or a walk. Watchers rather than watched, they

for example, is the sister of the farm's senior driver. These women are therefore not among those discussed below who experience extreme alienation and vulnerability during their contracts.

are able to maintain the unregulated rhythms that shape the workplace outside the harvest.

The work of *mapermanent* for much of the year comprises diverse tasks, carried out in small groups according to variable rhythms. This is because the permanent workforce represents a continually available source of adaptable labour, on call at any time. Grootplaas manages with a relatively small permanent workforce by keeping workers flexible. A builder might be taken on because construction work needs doing, with a view to employing the recruit as a farm labourer more generally. Most workers therefore sign contracts simply as 'general workers'. This workforce is divided into teams: Citrus (trees maintenance), Irrigation (pipes, et cetera), and the rather generic catch-all 'Lands' (from which harvest-time picking-gang supervisors are drawn). But workers can be moved around to suit labour demands, and a worker's level of responsibility is usually more important than his job description. Work days vary greatly in length, and from worker to worker. The flipside to this is that employees may be called to work at all hours: to remove a log stuck in one of the pumps in the Limpopo River late at night; to help fix a farmer's pick-up truck that has broken down in the game farm during recreational hunting; or to switch on an irrigation tap between beers on a Sunday afternoon.

It is not the contrast between harvest and other times that dictates the difference between this varied irregularity of permanent work and the relentless effort of seasonal work. Rather, such contrasts are a feature of employment category. As we have seen, many permanent employees occupy supervisory roles that allow or dictate variation in individual pace. Others from the core workforce – involved in irrigation, orchard maintenance, security, or, for the few permanently employed women, domestic work – remain largely unaffected by the harvest in their daily duties. Agricultural work regimes display similar contrasts to those found in industrial settings (see Parry 1999). Core employees labour according to a variable 'task orientation' (Thompson 1967), in which spikes in work are matched by long periods of rest. Seasonal labour, tightly coordinated at Grootplaas to process a huge volume of fruit within the tariff window of trade to Europe,[5] contrasts sharply with the task orientation and personal autonomy of core employees.

Seasonal workers, then, engage in tightly controlled, intensive, industrial-style labour. Appreciating the wider meaning of this contrast requires looking beyond work itself. As in Mollona's (2005) case of a Sheffield steel workshop, contrasting types of work create a deep sense of difference among workers as particular kinds of people. This sense of difference naturalises labour hierarchies and job categories, blurring

---

[5] During the northern hemisphere's summer months – citrus are winter fruits.

the lines between employment and life outside working hours. Roles in the labour process are only the beginning of the differences between modes of living at Grootplaas. 'Permanent' has up until now referred to an employment category characterised by open-ended contracts but, because housing is provided with employment, it is also shorthand for open-ended residence at the farm more generally. This is all the more true because of extremely low labour turnover among core employees: during the period of fieldwork, only three *mapermanent* left employment to my knowledge, and only two by choice (the third had been caught poaching).[6] Permanence is further a matter of workers' ability to establish a sense of rootedness through domestic congeniality, something that *mapermanent*'s work positions enable them to achieve. What is at stake here is how workers' roles in production – agricultural employment – affect their reproduction – maintaining the conditions for life in the compound. Turning to the Grootplaas labour compound reveals the broader meanings of workers' categorisation on the border farms.

Living in No Man's Land? Transience and Rootedness in the Compound

When farmers speak of 'the compound', they have in mind a single, uniform entity, characterised by reference to the race and/or culture of those who dwell in it. Spatial control according to colour has long been a characteristic feature of Southern African landscapes built around white settlers and capitalist production (especially Zimbabwe, South Africa, Zambia, and Namibia). Black living areas in towns and on mines and farms were built to control resident populations and were characterised by regularity, austerity, and lack of permanent rights of residence (see Gordon 1977; McNamara 1978, 1985; Moodie with Ndatshe 1994; Ginsberg 1996; Lee 2005). On farms, compounds – historically also known as *kraals* or farm villages – are widely understood to be the proper place for black sociality. Sometimes such areas were and are collections of worker-built, mud-and-pole accommodation – sites of far less thoroughgoing control and surveillance than mine compounds or townships. But on large farms like Grootplaas, owners have built brick housing with corrugated metal roofs and metal doors. Ironically, this is both the best farm-worker housing around and the accommodation that most clearly replicates the distinctive township and mine-compound layout, with its connotations of racial separation and utilitarian drabness. By contrast with the lush, private worlds of white farmhouses (see Chapter 2), Grootplaas's compound appears a bleak, regimented place. Employer control

---

[6] There were, however, deaths. Periodically, workers who are very sick go home on indefinite leave, and never return.

is implicit in the layout, where long rows of identical brick cells are easily surveyed and scanned by farmers, foremen, or police. The majority of Grootplaas accommodation takes the form of single rooms arranged in blocks of six, each with its own external door and a small window. Public showers – segregated by sex – are in urine-stained, roofless rooms, in which nozzles in long pipes along one wall release cold water from the farm's boreholes. Pit-latrines are in doorless, concrete cubicles in roofed but unlit buildings, making night-time visits challenging at best. For the most senior workers, by age and place in the work hierarchy, these conditions are somewhat improved by two-roomed, semi-detached houses, with outhouses containing private showers and flush toilets.

This depiction, with its emphasis on austerity, corresponds to the view taken by many seasonal workers who arrive at Grootplaas. New arrivals see primitive, uncomfortable, prison-like, alienating cells. However, belying the compound's apparent uniformity, permanent workers see it very differently: as a place of everyday domicile in which their lives are rooted. In a manner similar to apartheid-era South African township-dwellers (see Ginsberg 1996; Lee 2005), *mapermanent* assert a sense of belonging by adapting their housing, planting gardens, and reproducing familiar forms of domesticity. Established residents transform accommodation in ways that bind their personal histories into the fabric of the place. There are differences in the extent to which people are able – or indeed want – to establish sustainable lives in the compound. Exploring this variation offers a window into how different patterns of settlement and movement intersect at Grootplaas.

The Compound from a Seasonal Point of View

Seasonal workers' difference from *mapermanent* is made clear from the moment of their arrival in the compound. They are allocated bare rooms in a thirty-block grid known as the New Houses (see Figures 4.3–4.6) or a long barracks building called the Hostel (see Figure 4.7), their room-mates often chosen by an appointed permanent worker from the *mapermanent*-organised Housing Committee. Cells are overcrowded – the five Hostel rooms hold twenty to thirty people each. They will have been uninhabited for the six months since the previous harvest and are sometimes rat-infested. Writing in chalk or charcoal on the walls and floors bears witness to the cryptic history of previous occupants.

Both the New Houses and the Hostel are in a distinct area of the compound, an exposed slope illuminated by a huge floodlight. This illumination, lack of vegetation, and the unobstructed grid layout allow the observer to see straight from one end to the other down the rows of blocks. Although a few permanent workers have rooms here, at the end closest to the rest of the compound, much of the New Houses section

Figure 4.3. The Grootplaas labour compound (the New Houses).

Figure 4.4. The New Houses, quiet during working hours.

(where I also lived) is inhabited only during the harvest. Women who do not have full-time jobs can consequently retain rooms in this area during the period before or after harvest, when they work for the farm. They make a living as illegal beersellers, operating shebeens. During the harvest, the New Houses are transformed from a relatively quiet area into a bustle of people, cooking fires, and, at the weekend, loud parties.

Residential separation is far from complete. Those *mapermanent* residents of the other areas who want to drink, party, and find women, for example, frequent the New Houses. Because the shebeens, and most televisions recognised as being for public use, are in the New Houses, the area itself has the air of a free-for-all, in which loud music, gambling, and publicly drunken behaviour are common. The rest of the compound, however, remains tranquilly unaffected. The New Houses and the Hostel are seen by many who live in the other parts of the compound as loud, dirty, and a site of immorality. Unsurprisingly, some men who speak of the New Houses in this manner nevertheless go there for recreation. Such visits, however, contain an element of choice, in that they can escape back to the relative tranquillity of their own areas of the compound – permanently occupied, better kept, and made up of housing units adapted into homes, taking the edge off the prevailing architectural uniformity.

Despite such moments of mixing, the fact that seasonal workers reside in a distinct area of the compound itself reinforces their difference from *mapermanent.* It means that they have little reason to spend much time in the other areas unless they know residents sufficiently well to visit them at their houses. The spatial organisation of the compound reflects a wider experience shared across much of the seasonal workforce – limited integration in social networks dominated by permanent residents. The New Houses area itself is easy terrain for border patrols to run down so-called 'illegals': it is packed with the seasonally employed, almost all of whom cook outdoors on fires (Figure 4.5), and it is an open, unobscured stretch with a regimented, easily navigable grid of uniform blocks.

There are, it must be said, differences among seasonal recruits' experiences of the compound. Arrivals with urban backgrounds, whose descent into farm employment because of hyperinflation and economic contraction at home is experienced as a form of degradation, see the compound as dirty, noisy, and alienating. Those with contrasting backgrounds see particular advantages: those from rural backgrounds appreciate electricity in the rooms and easily accessible boreholes for water; some of the young find opportunities, in the noise and overcrowding, for parties and sexual adventures. A crucial fault line in experiences of the compound is between established residents and new arrivals. This does not map exactly onto employment categories, however. Some (especially female) seasonal workers are the partners or relatives of permanent employees and inhabit the compound – including the New Houses – during non-harvest

Figure 4.5. Seasonal workers sit around a fire in the New Houses.

periods in a variety of domestic arrangements. They are well known at Grootplaas, receive legal documentation most quickly because of their connections, and have access to better housing. Contrasting experiences of permanence and seasonality, and their spatial connotations, are therefore not simple reflections of the labour hierarchy. This point is explored in the next chapter. Here, it is relevant to note that such opposed and contrasting ways of living at Grootplaas are shaped by the complex web of relationships in the workforce.

Despite the different social positions of seasonal workers, what the majority share is a set of adverse living conditions. Unlike better-established residents, the way they are housed leaves them with little ability to shape their environment. This is as much a product of their precarious positions and transience in the area as it is a product of the compound's architecture. At the beginning of the harvest seasonal employees, concerned to evade deportation, often avoid their rooms altogether.[7] And throughout the harvest most seasonal workers, having arrived recently and with no investment in compound life, have few comforts in their rooms. Their short time at the farm and their insecurity discourage any significant attempts to adapt accommodation to create congenial living

[7] See above, p. 151.

Figure 4.6. Inside a New Houses room.

arrangements. Most pickers' rooms are spartan, with cardboard on the floor to sleep on, some food, perhaps one or two spare items of clothing, and a piece of wire rigged up as a clothes line between two walls (Figure 4.6). More extreme are living conditions in the newest housing, the Hostel (Figure 4.7). Residents – men in most such rooms – sleep twenty or thirty to a room, and complain of lice and sick people coughing in confined conditions. With little space inside, and no electric lighting, residents choose to sit outside around fires when they are not sleeping. Soldiers regularly move through the groups with torches, checking their farm IDs. This cramped block, a clear target for border patrols, epitomises the difference between the ways seasonal workers and permanent residents experience life at Grootplaas. It also strongly recalls the infamous mine labour compounds of South Africa's past.

The way seasonal workers live in the compound is especially central to their experience of Grootplaas because their day-to-day existence is largely confined to this distinct area and to the workplace. Otherwise, they are hemmed in by vulnerability born of their undocumented status. The farm is treated by the South African army and police as part of the 'border situation', as we saw in the previous chapter. Similarly restrictive is the fact that many seasonal workers never receive work permits – at the time of fieldwork this was largely the result of bureaucratic inefficiency

Figure 4.7. The Hostel.

in the South African Department of Home Affairs and the Zimbabwean border authorities. Harvest time attracts enormous police attention: the border farms suddenly house large numbers of undocumented Zimbabweans, who become easy targets for deportation.[8] Although Zimbabweans without work permits run the risk of being deported at any time during the year, in practice the aggressive police raids begin only with the harvest. Police vehicles, often with an army escort, move through the compound at unpredictable times, rounding people up. The afternoon after a weekend police raid, the compound would be deserted except for a few permanent employees drinking beer, its other inhabitants hiding in the bush. Farmers negotiate with the police and army. They secure agreements that identity cards produced by the farms will serve as proof of the holder's pending 'legalisation'. But it is some time before such *ad hoc* deals become known to police on the ground, and thus take effect. Further, the farm identity cards are not recognised as offering protection beyond the boundary of the farm in question. Although police attention tails off during the course of the harvest, the undocumented remain vulnerable.

---

[8] As noted in Chapter 1, deportations ceased in 2009, but resumed – albeit less zealously – in late 2011. See Chapter 1, footnote 22.

Because of their vulnerable legal status, it is difficult for undocumented seasonal employees to move around the border area. Walking to work in the morning along the border road, they risk being picked up by army patrols until they have some recognised form of documentation. And walking home from overtime work at the packshed in the thick darkness, workers avoid using torches in case it attracts soldiers. Particular work locations, such as the packshed, are areas safe from the threat of deportation. But the police enter the orchards – usually to demand oranges – and their arrival sends pickers sprinting through the trees. Always on the lookout for police raids, seasonal workers tend to confine their movements to shuttling between work and compound. While the deportation of workers is inconvenient for farmers, one result is an unassertive labour force whose everyday movement is sharply delimited, as mentioned earlier.

Transient vulnerability and austere accommodation, one handicap reinforcing the other, emphasise seasonal workers' positions as short-term units of labour. It is in contrast to this experience of seasonality – where workers remain vulnerable, easily controlled, and confined to designated spaces (work and compound) – that the lives of *mapermanent* need to be understood.

Adaptation and Rootedness Among Mapermanent

Mapermanent see Grootplaas as their home for their working lives. It is not that any of them intends to die and be buried at the farm. Retirement means returning to rural homes in Zimbabwe that have been developed gradually over the years, in the classic mould of regional labour migrants. After the end of waged employment, success for many will be defined in terms of a respectable life *kumusha* (ChiShona: at their rural homes). However, they invest – socially and materially – in their lives at Grootplaas. Some have not been back to Zimbabwe for years, and rarely even mention homes across the border.[9]

Mapermanent's sense of rootedness both reflects and is augmented by the fact that many have long personal histories of residence at Grootplaas and nearby farms. Indeed, the longest-standing residents have been living on the border farms since the days when other whites owned the land, before their bankruptcy. They can trace the boundaries between the old estates, before they were consolidated by the current farmers. One such long-standing farm dweller is Marula, the foreman, who was born on an estate down the road where his father was a foreman. His many children were born and grew up at Grootplaas or on neighbouring farms.

---

[9] Regarding migrants elsewhere in the region who, despite rarely visiting rural homes, nevertheless preserve an ideal of rural connection and retirement, see Bank (1999) on men in East London hostels.

Producing Permanence 123

Although they now live in a variety of places – including Johannesburg, Musina, and the former Venda homeland to the southeast – they visit regularly. Marula's teenage daughter, who boards with her grandmother in Musina while she attends high school, spends most weekends with her father. For some of Marula's children, the border farms remain their places of domicile. Marula's youngest son, a toddler, lives with him and his wife. A boy of six years, born to a different mother, stays at her home on a nearby estate, but spends a lot of time in the Grootplaas compound. An adult son, Mpho, is a permanent worker at Grootplaas.

Such personal histories at the farm mean that *mapermanent* experience a far greater sense of local attachment than might be assumed from a narrow focus on their employment. Like other longstanding farm dwellers, Marula remembers the construction of current dwellings, and the existence of previous compounds, now disappeared. He can pinpoint the place of his now-adult daughter Takalani's birth. At that time, in the 1980s, it was the site of compound housing; now, it is a patch of nondescript scrub on the edge of the Grootplaas football pitch. Another old resident – a longstanding friend of Marula's – recalls how, in the past, people would live in one compound, as a base, and work on different farms up and down the border, sometimes for food rather than cash wages.

The memories of permanent black residents are shared and maintained through naming practices. Among themselves, permanent workers have their own names for both their employers and the estates, each name encoding a history. Willem, one of the Grootplaas farmers, is known as Mpothe. What is the significance of the nickname held by workers to mean 'Hit'? One version of the meaning cites his history of violence towards workers, another his short temper with nosy police during the days of apartheid. Either way, his temperament is noted. The farmer who previously owned the Grootplaas land, Gert van Wyk, was known as Re a tseba, Northern Sotho for 'We know' – he would often underline his command of the language by announcing this phrase to workers. Compounds and areas of farms are known by the names of present or past owners: 'Paul Compound', 'Shala [Charles] Compound', and 'KK'. Grootplaas is still known as KhaRudi, after the son who left to operate the family's Kleinplaas venture but remains workers' favourite farmer. Other farmers and areas are known by names whose meaning has been lost, but which frame places and people in a parallel language to that of white landowners.[10] *Mapermanent* assert rootedness on the farms through their own local historical consciousness, their equivalent to the similarly motivated historical narratives of the white farmers (discussed in Chapter 2).

---

[10] See van Onselen 1976 for similar practices in early twentieth-century Southern Rhodesia.

Mapermanent assert a sense of belonging that goes beyond their employment. They do so not only through their shared histories – both told explicitly and encoded in names – but also by adapting their accommodation in the compound, mirroring the homes to which they aspire in rural Zimbabwe. Established residents of the compound develop their housing, accumulate furniture and other goods, and plant gardens, investing in their lives on the border. They do so despite the fact that they live on an employer's land, and will have to leave if they resign their jobs or if they are sacked. Such adaptation of and investment in precarious accommodation parallels accounts of other spaces in Southern Africa defined as being for black wage workers. Residents of mine compounds built furniture to improve their bleak accommodation (see Gordon 1977 for Namibia), while inheritors of so-called 'matchbox' housing in 1960s township areas like Soweto added flooring and ceilings, plastered walls, and planted gardens (Ginsberg 1996; see also Lee 2005). In both cases, they did this in spite of the fact that they lacked any security of tenure and could be ejected at a moment's notice. Indeed, in Soweto, they did so precisely to assert a more enduring right to stay: an expression of rootedness addressed both to other people and themselves. As at Grootplaas, continued residence depended on employment, itself at the discretion of white bosses. But at the farm, as in these other cases, adapting housing is an important way to achieve dignity, respectability, and a domestic life beyond such precariousness. Indeed, it claims a wider engagement with the place, beyond its status simply as a site of employment.

Where the housing of *mapermanent* appears uniform and barracks-like, then, these residents are in a position to adapt and personalise it. They are provided with an unfurnished room (or a two-room house for the most senior workers) in less bleak and windswept surroundings than New Houses occupants endure. But over time they build a bed inside, often from wooden forklift truck pallets topped with a sponge mattress. They obtain or build shelving and often decorate it with lace, to store and display crockery and other effects. Such displays of accumulated goods mirror demonstrations of established domesticity in houses in rural Zimbabwe. Practically all permanent residents have electric stoves in their rooms. Stoves make cooking straightforward and relatively quick. (Seasonal workers, in contrast, have no option but to gather wood and light fires outside in order to do their cooking.) Fridges are not uncommon and a minority of residents buy very large freezers to store beer and meat, which they later resell. The number of aerials towering above the brick blocks testifies to the wide ownership of televisions. Indeed, electrical goods are far easier to operate in the compound than they would be in rural Zimbabwe, where many areas lack connection to power grids and appliances require car batteries.

Some adaptations demonstrate both domestic propriety and success. Most residents rig up a wire between two walls, and hang a piece of cloth next to their beds. This way, when their doors are open or they entertain a visitor, the bed area remains out of sight. Screening off the bed allows residents to display their decorated shelves adorned with possessions, while distinguishing between degrees of privacy. Doing so follows the layout of homes in Zimbabwe, where houses are often built with multiple rooms, one a living room replete with decoration and display. In the compound, both decoration and bed screening are important because people sit outside their houses – metal rooves without ceilings make for stifling heat when it is sunny – and leave their doors open most of the time.

In their endeavours to adapt their housing, residents rely on one another. When personnel manager Michael wanted a concrete step outside his house to keep the rain away from his door, he enlisted the help of the farm's builders, who used spare cement from the new Hostel accommodation. Others depend on access to the pile of scrap by the farm's workshop, building furniture for themselves and for friends. They also connect metal poles end to end, creating the tall television aerials that are ubiquitous in the compound. Some projects, however, are more elaborate. A three-legged metal stool took one worker several days to complete. Beautifully crafted, it was an object of personal pride and a gift for his neighbour. Some adaptations of accommodation, therefore, deepen a sense of rootedness in the compound because they attest to the dense web of relationships among permanent residents.

Similar improvements are evident in the space outside the house (see Figure 4.8). Some residents build yards with brick and/or mud walls, or plant hedges. Or they build *stoeps* (verandas) to demarcate the ground outside their doors, using concrete or bricks left over from the farm's building projects.[11] A few have even planted trees. Most permanent male residents have vegetable gardens, either by their houses in the case of senior workers, or on the edges of the compound. Not only do these gardens provide edible green plants known as *muriwo* or *muroho* (ChiShona and TshiVenda respectively), a source of relish for *sadza* or *vhuswa* (ChiShona and TshiVenda: maize-meal porridge); they also underline the permanence of these residents in the farm's landscape. Just as everyone knows each resident's room, so what appear to be large swathes of vegetable patch are in fact several well-marked gardens whose individual

---

[11] A few residents invest yet further in their compound accommodation. Marula, the foreman, as the longest-serving and most securely established black worker, has built an entire compound of his own. His home is in the centre of the larger compound, and he holds court there, with other senior men, to judge disagreements between residents. His specific case is discussed in Chapter 6.

Figure 4.8. *Mapermanent* in a yard constructed outside a room.

ownership is widely known. Such assertions of belonging among *mapermanent* rely on the fact that people know a great deal about one another and their business: not only where a particular permanent resident lives, but also his or her occupation at the farm, and a web of stories and rumours.

The permanently employed make homes out of their accommodation, adapting it and shaping their apparently rigid environment. Unlike many seasonal recruits, long-term residence makes their relationship with the farm one which involves a great deal more than mere labour. Transient seasonal workers do not do this. They are not around for long enough, live too precariously, and often have no wish to become better established. Instead, they are eager to move on and away from this inhospitable setting. These mobile, short-term workers would appear, from one point of view, to be ideal as dispensable units of labour – their contracts clearly limited, their movements regulated, and their relationship to the farm tenuous.

However, this contrast is too simple, because it assumes that *mapermanent* establish themselves at Grootplaas in isolation from the more transient population around them. In fact, *mapermanent* draw transient people working on different rhythms of settlement and movement into their own projects of rootedness. They do this in two ways. They initiate relationships with mobile women in projects of domesticity. And they maintain the border farming area as a dense community that includes the soldiers, even though the latter are apparently there to enforce the border in a strictly impersonal manner. In the next section, I describe each of these spheres of *mapermanent* influence in turn.

Rooting Relationships

Mapermanent, as secure, waged workers just across the border from Zimbabwe, represent stability to those passing through: seasonal recruits, other migrants, and soldiers doing their time on the border. The ways they draw transient people into their own lives is usefully conceptualised as 'place making', in Feuchtwang's definition of 'the centring and marking of a place by the actions and constructions of people tracing salient parts of their daily lives as a homing point in their trajectories' (2004: 10). Feuchtwang emphasises the *gathering* quality of place: the orientation of different locations and movements around a focal point. *Mapermanent* draw people into their lives for their own reasons, but this also 'gathers' various residents of the border area into communities, however provisional, centred on the compounds. This perspective takes us beyond the way Grootplaas residents engage with compound accommodation itself, to consider how the farm represents an important spatial centre of gravity, with *mapermanent* at its heart.

Shifting Domesticity

Unlike seasonal recruits, *mapermanent* have their own housing, work permits, and stable incomes. Apart from the immediate benefits, they are also consequently able to attract women as live-in partners. Women come to the farm to seek employment, either in the seasonal workforce or caring for the children of *mapermanent*. Seen in purely instrumental terms, from the perspective of women, especially young women, influential men on the farms appear to have clear prospects in what might otherwise be a desperate situation. At the same time, from the male *mapermanent* perspective, it is through relationships with women who are moving through the area that they are able to replicate homely domestic arrangements. Permanence, in the sense of rootedness, is asserted and experienced as congenial domesticity. Long-term male workers establish lives in ways that cohere with their gendered expectations, through attracting the labour of a floating population of young women. In some cases they establish farm-based 'marriages'. The tension, between women's concerns about their material insecurity and men's concerns to create an agreeable home environment, reflects the enormous inequality between them, in terms of access to income and accommodation. Like South African migrant hostels (see Ramphele 1993), the Grootplaas compound is highly gendered, with established men claiming the space as their own. Many women, in coming to the farm, have taken decisions relating to their existing roles as wives, daughters, and mothers of young children; many, in turn, take on roles in relation to men in the compound, as sexual and domestic partners.[12] But the situation at Grootplaas reveals more than this. The very rootedness of permanent male residents is *constituted* by transient relationships with women, even as they reinforce the insecurity of these women.

When I arrived at Grootplaas at the end of 2006, my overwhelming impression, contradicting what I had expected in a border setting, was one of stability. Michael, the personnel manager, had offered me residence in his house in the compound and I soon found myself to be a member of what looked a lot like a stable household unit. All members were Zimbabwean, but they seemed to have made a home out of the barracks-like compound accommodation on this South African farm. Michael lived in the main house with his partner, Purity. Three young women lived in a mud-and-brick room in Michael's yard, in return for cooking and cleaning. Michael and Purity expected to bring their young daughter, Lindsay, to the farm after Christmas. One of the young women had a baby. There were always people around the house, cleaning, resting in the yard, watching television, or preparing three daily meals.

It soon became clear that I had arrived in a household, and one situated in an established community in which there were ties of a more or less permanent kind between neighbours. But my initial impression of stable domestic life, itself contradicting the type of shifting existence I had initially expected, proved in turn to be somewhat mistaken, at least in the rather static form that my imagination conferred on it. When Michael returned after Christmas 2006, he came without Purity, who stayed to look after their sick child in Zimbabwe and whom he would soon abandon in favour of a new, pregnant partner, Holly, whom he had met at the farm. The three young women did not re-establish themselves at Michael's after their Christmas visit home to Zimbabwe. One, Wonder, returned to the compound for a while but stayed with Michael's neighbour, Gregory; she rarely came to Michael's yard. Suddenly Michael and I were living alone, eating far fewer cooked meals and more dry, white bread with soft drinks. For Michael, as for many permanent workers, 'household'-like structures depend on domestic arrangements that involve mobile women. Arrangements were as fleeting and impermanent as the presence of these women themselves, something I found extremely alienating after the apparent conviviality of compound life.

---

[12] As in the case described by Elder (2003).

Michael soon responded to the change, complaining that, with a long work day, he did not have time to cook. At his suggestion, we employed a young woman to work, as he had done the previous year. She cooked, cleaned his house and my room, and washed clothes, in return for meals, shelter, and pay. She soon moved on. Later, during the harvest, Michael's sister Pula and niece Lovely came to work. Pula needed an income to bring up her son, a child whom she had left in Zimbabwe. She had cattle, but hyperinflation had meant that there was little point in selling them. Lovely had passed four O-levels,[13] and hoped to 'expand' – to take more subjects. The money from farm work was intended to pay for home-based education. This was because Zimbabwe's school system was disintegrating, as teachers' salaries became worthless and parents were unable to pay school fees. Michael ensured that they were employed at the farm. They meanwhile fulfilled all domestic duties during the hours after work, this time without pay. And they continued to do so even when they acquired their own room, until they left the farm after the harvest. By then Holly had given birth, was back on her feet, and was keeping Michael's house again.

It is important to note that not all domestic arrangements at Grootplaas are as transient as those just described. Indeed, for Michael, the presence of Holly represented the beginning of a steadier period. Many permanent workers, some married at home, have relationships at the farm. And some of these develop into permanent farm-based arrangements, which endure sufficiently for couples to have children. Such farm partners are not taken home to Zimbabwe. At Christmas, for example, each may go back to a separate marriage and family for a few days. Nevertheless, such relationships render Grootplaas an important place – it is the only context in which they have permanence. Indeed, an informant warned me when I was conducting an interview that asking about marital status was sensitive for precisely this reason. What I might seem to be probing, she told me, was whether the respondent was *really* married. Whether, in other words, the marriage was ratified and not 'merely' something that was confined to the farm area.

Such 'farm marriages' are far from simply domestic 'arrangements'. At their most stable, they are the basis of families, once again giving greater meaning to 'permanence' at the farm than simply steady employment. The relationship between Norman, the farm's senior driver, and Joyce, another permanent employee, is an example. Norman's wife at home is Sarah, who visits a few times a year. At the same time, he has a well-established, openly declared relationship with Joyce. Together they have a five-year-old son, Sam. Although Norman keeps his own house, where Sarah stays when she visits, *he* usually stays at Joyce's. The couple have

---

[13] Basic level high school exams. The stage before A-level.

cultivated a comfortable homeliness, and have adapted her room with a sun-shelter outside the door and an old seat from a bus so that Norman and his friends can watch the football on television there. Norman and Joyce represent, for their friends among the *mapermanent*, a point of domestic stability around which they can congregate in their leisure time.

Norman and Joyce have a relationship that resembles a common practice in Zimbabwean townships and on farms, known in ChiShona as *mapoto* (see Barnes 1999). *Mapoto* – literally 'pots' – describes domestic arrangements without formal marriage (ratified through *roora* [bridewealth], church wedding, or state registry). In *mapoto*, women move in with men and receive everyday support from them – food and shelter especially – in return for domestic labour ('pots') and sex. Such arrangements developed as a response to situations where men had almost exclusive access to housing, but often lived at a distance from their wives. Women in the arrangements bore the brunt of moral condemnation, having disrupted both patrilineage reproduction – as men and their families had no claim to the children – and notions of respectable propriety. At Grootplaas, as in Zimbabwean townships and farm compounds, male access to housing and their stable wages shape women's options. But unlike in those settings, the more settled of these relationships are not condemned, but rather seen as permanent within the farm context. There are few actual marriages at Grootplaas in which both partners are present. In their absence, the better-established farm relationships are seen as positively respectable, and women in these partnerships stay at the farm all year round.

However, such sedentary domesticity is the experience of a minority of compound residents, often the most senior of the *mapermanent*. In fact, there is a constantly shifting population of mobile women, more or less attached to settled residents. Permanent workers' wives sometimes come to visit, and while they stay at the farm they look after their husbands and their houses. Other women pass through the compound, either heading south or crossing the border to earn a bit of money at the border farms before going back home. This needs to be understood in its particular local context. The area across the border in Zimbabwe is particularly marginal and under-resourced, home to a small, TshiVenda-dominated minority with little access to employment (Mate 2005). The area is further prone to drought, and the only real alternatives for those seeking waged work are the sugar plantations of Hippo Valley, Triangle, and Mukwasini to their north, in southern Zimbabwe (Lincoln and Maririke 2000: 43–4). With few options besides connections to the South African border farms, many women come for agricultural employment but then follow other opportunities for livelihoods.

This perspective was impressed upon me by Margaret, whom I knew as Granny. Margaret, described in greater detail in the next chapter, was the wife of a security guard at Grootplaas, before she and her husband moved to Johannesburg in 2007. She was 47 at the time, had A-levels and qualifications in teaching and social work, and had worked as a teacher before coming to the farm. At Grootplaas, she had gathered information for a non-profit organisation[14] about the lives of women, and she was keen that I hear of their situations. During the harvest, she accompanied me in conversations with young women, to whom I would have had little access had I not been with her. In the compound, we spoke to teenage girls as they babysat workers' children during the daytime. In the orchards, we spoke to female pickers, a small minority of the picking workforce, who were grouped into separate teams from men. What became clear was how limited many women's options were, as they sought a means of sustenance.

Some young women come to the farm from close by across the border to look after children during the harvest. Female seasonal workers pay girls, sometimes still in their early teens, to look after their children during the day for a pittance (as little as R150[15] per month). Such girls may live with a relative, who decides how much they should have for themselves and how much they should remit. They may be the only source of cash for their parents. Other young women seek formal work but find only shelter with a permanent worker in return for cooking and cleaning. Although some women have connections to friends or relatives on the farms, regular sojourns at the farms should not necessarily be taken as a sign of attachment. Women are extremely vulnerable and often find themselves without redress in cases of physical and sexual violence. Among themselves, men speak of persuading them 'by force' to engage in sexual intercourse. Ulicki and Crush report, of women migrating from Lesotho to South African farms, that 'fear and loathing are everywhere, bitterness prevails' (2000: 76). Women in this case, as on the Zimbabwean-South African border, return more because of preference for the known over the unknown than out of any enduring rootedness. While migrating southwards into South Africa is always a possibility, it is a radical step into a world of which they have little knowledge, in comparison to stints on the border farms a few kilometres from home.

In such precarious circumstances, it is common for women, unemployed or employed, to establish domestic relationships with resident men. Doing so can, among other things, ensure a period of material security and even connections to influential figures like Marula, the foreman. Though often motivated by real affection, and occasionally leading to enduring unions, these relationships are shaped by their wider context of need, exchange, and distribution. As in other places built around

---

[14] At which her cousin worked. See next chapter.
[15] Approximately £10 at the time of fieldwork.

resident male workforces, women's lack of secure access to income necessitates a degree of 'economic realism', an awareness that there is a transactional dimension to relationships (Vaughan 2010: 22). Like *mapoto*, relationships involve material support and shelter for women in return for domestic labour and sexual access. For women, one danger is becoming pregnant. Often men do not take responsibility for their children. A child represents another mouth to feed, tying women into even greater dependence on future farm employment and further supportive relationships with wage-earning men.

Unattached young women seeking relationships with men face not only these dangers, but also the condemnation of other women. Regarded as threatening respectability and existing relationships, they are sometimes likened to prostitutes. It is notable, for example, that while Sarah is unhappy about Joyce, she reserves her scathing public comments for *younger* women who come to 'stir up trouble', when she chats with women of her own generation. On one occasion at the end of the harvest, she and others complained that these younger women were waiting for seasonally employed wives to leave for home in Zimbabwe, so that they could work on their husbands. Some men are indeed on the lookout, and some sex is explicitly transactional, for money or payment in kind (see Addison 2013; Hall 2013). Michael, for example, during a period without women in the household, turned to finding other possibilities. One morning I turned up at the house to find a young woman cleaning and making breakfast. I had never met her before. In the context of women's limited options and precarious lives on the border, men's permanent employment – its stability and power in the farm setting – enables particular ways of living at the farm. In this case, Michael had reportedly promised the woman a seasonal job in the packshed – a use of position that is by no means unusual.

For many women, domestic labour in the compound defines the very rhythm of the day. By the time the sun warms the air around seven o'clock, unwaged female residents are sweeping dust out of their rooms. Many have either already prepared packed lunches for waged men, or will begin cooking for the midday meal in the late morning. And between these meals there are further tasks: hand-washing and hanging clothes, and washing dishes and pots. Afternoons offer a period of leisure, but by around four o'clock women begin disappearing to shower and then cook dinner in time for the end of the farm's working day at five. All of this follows highly gendered notions of appropriate work that extend all the way into the farm's waged economy. The only women permanently employed at Grootplaas are domestic workers in the offices and in white residents' houses, and two keepers in the compound shop. A further twenty or so women remain on the farm throughout the year as 'semi-permanent' employees. Often from close to the border, and with a history of kinship connections on the farms, they only work full-time during the harvest but are paid by the farmer for part-time casual work throughout the year. One regular task in such waged employment is cleaning the compound – picking up rubbish and sweeping the ground. Even waged employment on the farm, therefore, reflects assumptions about the femininity of domestic labour.[16]

Understanding the different ways of living in the compound requires appreciating how men's place is made in part through female domestic work. Permanently employed men are better able to secure access to women, for sex and domestic arrangements, than transient seasonal workers grouped into cramped shared rooms. It is through relationships with women – many of whom are highly mobile – that permanently settled men are able to achieve living conditions that approximate those of home, in which women clean houses and clothes, and prepare food. Historically, workforce cohesion and embeddedness on South Africa's and Zimbabwe's white farms have relied on domestic arrangements within the workforce (see, for example, van Onselen 1992; Waldman 1996). Here, such arrangements depend on the fleeting appearance of women in male workers' compound houses.

Such domestic arrangements contrast with the lives of many seasonal workers. The latter's residence at Grootplaas is in all-male rooms, outside which they cook for themselves on fires. Their experiences recall the much-maligned regional history of labour migrancy, with its restriction of movement and residence, and the racialised control of space. While young men and women in the seasonal workforce do establish sexual relationships, these are generally fleeting, and not built upon domestic arrangements. Transient men, housed in groups, lack both the stable incomes and the relatively comfortable and private accommodation necessary when establishing the kinds of relationships that ensure female domestic labour.

Localising the Military

In similar vein, but in a different register, Grootplaas residents build provisionally stable living arrangements in collusion with soldiers on border duty. As noted at the beginning of this chapter, soldiers establish connections with members of workforces on the border farms. They do so despite the fact that they are regularly rotated between different border garrisons, making it difficult for them to become well embedded in

---

[16] Despite the parallels between domestic work for wages and domestic work in relationships, the former is certainly regarded as employment, while I encountered no evidence that women see their live-in arrangements with *mapermanent* in an employment idiom. As openly remunerated work, the babysitting and cleaning described above is seen as employment of sorts.

the social lives of the compounds. They also overcome other obstacles – being kept at a distance by some compound residents; and lacking the linguistic proficiency necessary to communicate (often they have been recruited in different and distant parts of South Africa). They are able to integrate to some degree because farm residents see them as a necessary fact of life, even speaking about them with sympathy as men alone on the border, assigned a thankless task. Soldiers, in turn, engage with farm residents in a sufficiently congenial manner to impede impersonal border regulation. It is common for farm residents to cross into Zimbabwe for the afternoon to find *Chibuku* beer ('traditional' beer, commercially produced in Zimbabwe) or to pass through the fence at the army guard post to go fishing in the river. They merely let the soldiers know when they intend to come back.

Soldiers, rotating among different garrisons every few weeks, become provisional members of farm communities. One way their role in the border populations is made explicit is as keepers of the peace. On one occasion, for example, when thieves were found within the Grootplaas compound, residents handed them over to the soldiers after beating them. The soldiers are drawn into the settled lives of border farm dwellers: people empathise with them and they are co-opted into dispute resolution; they also offer a degree of everyday security.

At the peak of the Zimbabwean crisis, this fragile relationship was disrupted during the harvest, when the army had to prove its worth by arresting large numbers of undocumented compound residents, often acting as escorts for the police. At this time, soldiers became less visible at the shebeens. However, they did not stop coming to drink or look for women in the compounds. Neither did farm residents entirely lose their sympathy towards soldiers. They saw them as having a better local understanding – including respect for the farms' informal employment card systems – and greater empathy with residents than the police. One informant told me that people at Grootplaas were not afraid of the soldiers because it was their job to guard the fence, not arrest people (although my own observation showed that they also do the latter). The police patrols came to arrest, whereas she had heard soldiers tell people: 'We understand things are tough that side [Zimbabwe] and you have to come here, but please don't cut the fence – otherwise, we get in trouble for not doing our job and have to chase you.' Soldiers and farm residents have to 'get along and not get each other into trouble', she concluded. The police, meanwhile – strangers on the farms and known for aggressive behaviour when rounding people up – received no such sympathy.

Many seasonal workers view soldiers with greater trepidation than do more established residents. The provisional stability created on the farms, through everyday cooperation between workers and soldiers, leaves recent arrivals – undocumented and afraid of deportation – on the margins. *Mapermanent* have employment permits. Unlike their seasonal counterparts, who often never receive their delayed papers, *mapermanent* are members of a small, indispensable workforce. Their papers are renewed every six months. This difference of legal status has important, wider implications. A greater familiarity with the area means that the better-established and better-documented do not react to the police or army presence, and are consequently rarely even asked for papers. Seasonal workers often run as soon as they see the *bakkies* (pick-up trucks) belonging to police or soldiers. Their lives lack much of the everyday predictability that permanent workers are able to establish on the border.

The inequality between *mapermanent* and more marginal residents of Grootplaas is clearest when soldiers are explicitly brought in to manage disputes. Like many other aspects of life at Grootplaas, the benefits accrue disproportionately to permanent workers. From the perspectives of many soldiers, permanently employed men have roots in the area, while they themselves are just visitors waiting to leave. *Mapermanent* act as gatekeepers to soldiers, buy beer for them, and are often able to choose exactly which conflicts are mediated and how they are presented. This offers women and those without connections few options for redress. In one case, a woman had been attacked by a senior permanent male worker. The dispute was brought to the soldiers, represented as one between the culprit and the woman's boyfriend. This was not seen as legitimate by compound women themselves. However, the two men were able to lend weight to their version of justice by invoking the power of uniformed, armed state officials.[17]

At one extreme, therefore, male permanent workers enjoy secure, congenial circumstances. At the other, new seasonal recruits and women remain at the whim of farm and state authorities. The contrast drawn at the beginning of this chapter – between the vulnerability and the security of farm workers – is part of an all-pervading distinction between *mapermanent* and Grootplaas's less-rooted population of seasonal labourers and unwaged, mobile women. The picture this paints is of *mapermanent* as a local aristocracy of labour, not a common way of describing farm workers in South Africa or Zimbabwe. This is true in the short term, reflected in day-to-day workforce and compound dynamics. *Mapermanent* enjoy forms of security and comfort denied to many of those around them. However, as this book's conclusion will demonstrate, the farms themselves, and therefore workers' homes, now have an uncertain future.

---

[17] It is important to note here that senior workers' impunity is tacitly guaranteed by the white farmers themselves, as I explain in Chapter 6. Though absent from discussions in the compound, farmers back their core employees by refusing to dismiss them in cases of abuse; such workers are seen as too important for production.

Conclusion

For farmers, flexibility and adaptability are crucial (see Chapter 2). As Chapter 6 will describe, many farmers consequently downplay a version of agriculture in which they have enduring paternalist responsibilities towards workers. They attempt to negotiate relationships with their employees through a shift to the language of markets.

But workers see matters differently. This chapter has explored how, unlike seasonal recruits, permanent workers' lives on the farm transcend mere waged employment. Rootedness is what permanence is all about, and *mapermanent* achieve it by drawing women and soldiers into their lives. They see the farms as home – a provisional home, different from the *musha* (ChiShona: rural home) or equivalent to which many will return when they retire, but home nonetheless.

This is a varied and complex picture. Schematically speaking, *mapermanent* are the most secure Grootplaas residents, while recent, unconnected seasonal arrivals and young women are the least secure. In between are regular, though sometimes unemployed, visitors to the compound, as well as better-connected seasonal workers. Structuring this inequality are the different kinds of employment experienced by the workforce. Differences of employment category take on further meaning in non-work areas of life, forming the basis of further kinds of inequality: between men and women, between those familiar with the border setting and newcomers, and between the well-connected and those without support. What all of these differences have in common, however, is that they are shaped by the ways *mapermanent* use their positions to maintain their own everyday stability.

This chapter's account of the place of *mapermanent* at the heart of the Grootplaas working population lays the groundwork for what follows. I turn now to a direct consideration of how structured social arrangements at Grootplaas intersect with large-scale changes and forms of fragmentation on both sides of the border. The next chapter focuses on male seasonal recruits and their experiences of displacement from Zimbabwe. It examines how they recreate their varied class and ethnic backgrounds at Grootplaas, through divergent forms of masculinity that are constituted in the work process and in compound life.

5 Reimagining Men
Middle-Class Farm Workers and the Zimbabwean Crisis

Introduction

When the harvest begins on the South African-Zimbabwean border, crowds of Zimbabweans appear at farm offices, hoping to be recruited (see the Introduction's opening vignette and Figure 5.1). Many have travelled by bus from other parts of Zimbabwe to the border town of Beitbridge. There, they hear of employment at the border farms. They make their way to the farms by *kombi* (minibus) taxi, but some are robbed and left stranded en route, and complete their journeys on foot. Others, walking southwards in search of work that will be remunerated in South African rands, stumble upon the farms. All cross the dry Limpopo River and climb through South Africa's boundary fence, evading soldiers, and border gangs (*magumaguma*) who are known to rob and rape.

Torn clothes and broken shoes from long journeys give the crowd a look of uniformity. But there are important differences among Zimbabwean farm workers – of ethnicity, class history, sex, and age. So while objective conditions may appear similar, Zimbabweans perceive their own place on South African farms and their participation in farm work in divergent ways. Zimbabwe's recent crisis has led countless Zimbabweans to seek unfamiliar modes of employment outside the country. For many, farm work represents sharp downward status mobility. Being well educated, as many of them are, makes it particularly demeaning to undertake what in Zimbabwe has long been a marginalised, low-status form of employment (Rutherford 2001). And, for some, engaging in such work *as migrants* presents further challenges to notions of respectability.

However, Grootplaas residents' self-understandings *at the farm* do not simply reflect their personal histories. As they are allocated rooms in the compound (see previous chapter), and as they encounter long days of work in the orchards or packshed, their backgrounds come to take on new significance. The markers of previous lives have meaning in farm-specific situations. Those who see farm work as a sharp drop in status – a sign of failure – are often keen to maintain a strong sense of distinction from the wider farm population. Given a historically powerful bourgeois

138   Zimbabwe's Migrants and South Africa's Border Farms

Figure 5.1. When the harvest begins: job seekers are kept out of the workshop yard by security guards and supervisors.

ideal in Zimbabwe (West 2002), it is perhaps unsurprising that a crucial way in which many farm residents make sense of their place at the farm is through idioms of class. This raises the question: *what happens to class consciousness in conditions of displacement?* How is it imagined and enacted? How do uprooted people strive to maintain a vestige of their own status, despite loss of livelihoods and social milieus? By exploring these themes, this chapter examines how experiences of displacement intersect with arrangements and hierarchies on the farm. It focuses on male pickers, for reasons of access. As I explained in the introduction, my opportunities for research with men were far greater than those with women.

Male pickers spend most daylight hours filling trailers with fruit. Picking in teams generates a form of camaraderie central in lending some unity to the experience of the majority. Meanwhile, those who see themselves as middle-class present the loud, sexualised banter of the orchards as everything they are not. On the one hand, a model of masculinity is constituted in the workplace; on the other, a minority assert their distance from this, and from the wider masculinity enacted on the farm, which workplace banter reflects. In the process, ethnic difference may also be mobilised. Some Shona and Ndebele arrivals characterise what they find on the farm as specific to the Venda, casting them as uneducated, as crude relics of colonialism, in contrast to their own perceived sophistication. They draw on stereotyped ethnic histories, but assert differences that map onto idioms of class.

Scholarly considerations of masculinity in Southern Africa have followed a dominant South African polarity between rural black areas and white production centres. Morrell remarks that 'the major configurations of masculinity which emerged as the twentieth century wore on were shaped by two major experiences and traditions . . . the workplace, primarily the mines [and] . . . rural life' (2001: 13). As discussed in the introduction, male migrancy has long been a dominant trope in analyses and imaginaries of Southern African labour, with especial focus on mining. Moodie with Ndatshe (1994) underline the importance of the wider conditions of migration for understanding migrant masculinity on mines. Their dual consideration of the migration and work contexts is instructive. At Grootplaas, both the rough-and-ready, dominant workplace masculinity and the middle-class response reflect experiences of displacement. Meanwhile, work itself is fundamental, with both ways of being male performed in the crucible of the picking process. But the situation at Grootplaas is crucially unlike classic cases of Southern African labour migration, such as that of the miners about whom Moodie with Ndatshe write. Dissenters from the dominant, team-based dynamics of picking work, in building understandings of their displacement, look to a Zimbabwean ideal of respectability that developed in explicit contrast to the stereotype of the unskilled male labour migrant (West 2002) – even though unskilled labour migrants are precisely what they have become on the farm.

The previous chapter demonstrated how differences between *mapermanent* and more transient members of the Grootplaas working population pervade both work and life in the compound. This chapter builds on that account with a more nuanced focus on seasonal workers. What it reveals is that seasonal workers, although very diverse, are polarised according to models of class, and these in turn are mediated through those of masculinity and ethnicity. Models of class, ethnicity, and masculinity reflect different reactions to Zimbabwean displacement, imagined and enacted in the farm setting.

Class, Displaced

To remind the reader, between April and September, Grootplaas employs up to 460 Zimbabwean seasonal fruit pickers and packshed workers to harvest the grapefruit and orange crops. High labour turnover means that the farm continues to recruit throughout the harvest. The workforce is spatially differentiated by gender. Known as *ma-cutters*, most pickers in the orchards are men. Most packshed workers are women. This is because grading and packing on the packshed conveyors is considered by farmers and workers to be women's work, while men make faster pickers. Indeed, the few female picking teams (and at some level women in general) are employed for reasons that cannot be reduced to efficient production. As one farmer explained, they keep gender ratios more even, ensuring that the compound is not simply full of men with unsatisfied sexual frustrations. This consideration is what underlies farmers' tacit approval of women's presence in the compound.

Seasonal work plays very different roles in different workers' lives, reflecting different responses to the Zimbabwean crisis. Luck, inclination, and timing are often all that separate people passing through from those who stay to work for a while. Some seasonal workers are strangers to the farm, some are not. Some intend to work the season and leave early, drawn by South African towns and cities with their less aggressive policing and their fabled opportunities. Others intend to pass through, but find work and stay; others still prefer to be closer to Zimbabwe and home, and come to work the harvest year after year, or move along the border farms hoping for employment.

Furthermore, plans change, depending on logistical considerations, goals (earning to send home groceries is a common preoccupation; there are also specific projects like building a house), and experiences at the farm. One significant factor causing such shifts in strategy is the limited amount that workers actually manage to save. Pay for pickers is low and, for men, entertaining girlfriends and buying beer can be expensive. Another factor is that Zimbabweans who come to work the harvest are unwilling to return home without something to show for their sojourn in South Africa. Having earned less money than expected, they start to make new plans on the spot. A third factor is that those (disproportionately young men) who hope to find work in Johannesburg often come back to Grootplaas after failing to secure employment. Far from the image many south-bound migrants have of Johannesburg as a place of opportunity, many in fact end up sleeping in the corridors of the city's Central Methodist Church. The building has been made available as a kind of reception centre for destitute arrivals, especially from Zimbabwe (see Introduction, p. 6, footnote 3). Such difficulties making ends meet in Johannesburg mean that, in the months following the end of the harvest, a steady stream of former seasonal workers arrives back in the compound hoping for bits of work, though by this time of year there are no openings. Many members of this floating labour reserve will return to the farms the following April to work another harvest. Despite the enormous diversity of backgrounds and recent migratory experiences among arrivals at Grootplaas, however, seasonal workers do polarise along lines of class aspiration to some extent.

For a significant number of residents at Grootplaas, both permanent and seasonal, working and/or living at Grootplaas is something of a disappointment. For some, farm work represents dramatic occupational status decline; for others, it is the result of failure to achieve social mobility promised by earlier successes at school. Their circumstances on the farm make for an arresting contrast with their aspirations, these in turn reflecting a middle-class ideal. Self-understandings take a variety of forms. They are not only shaped by giving meaning to personal *backgrounds* – age, sex, education, and occupational history – but also constrained by individuals' positions *at the farm*. Before turning to this, we must first appreciate how models of class developed in Zimbabwe.

Under settler rule, the existence of an African middle class did not fit a settler-conceived social order: blacks were assumed to be peasants or unskilled, low-paid workers. As elsewhere in Southern Africa, many members of the urban black population had come as labour migrants, and were expected to return home after their factory/mine contracts had ended. 'The state's interest in the development of a small African middle class' nevertheless left aspirations largely frustrated by racialised social stratification (Barnes 1999: 93–4). A class-conscious black elite framed its protests and claims within the idiom of a colour-blind 'civilisation' or 'respectability'. They stressed 'bourgeois domesticity', which included women's self-presentation as perfect home makers; (unsuccessful) efforts to acquire freehold housing, away from municipal townships; temperance; and conspicuous church 'white weddings'. Of particular importance was extensive (primarily mission) education (West 2002).

The Samkange family, about whom Terence Ranger (1995) has written, were prominent in this elite, not only in emerging African nationalist politics but also in redefining notions of propriety, family, and gender. Central to their position was Thompson and Grace Samkange's strong marital partnership, and their 'modernising' emphasis on education, including university, and church participation. Grace, in particular, became an important role model through her prominent position in the

Ruwadzano, as the African Women's Prayer Union became known in Rhodesia (ibid.: 40).

> Great leaders of the Ruwadzano like Grace were famous and honoured figures amongst middle-class Africans in Southern Rhodesia. It was these women who above all sustained the Christian monogamy which was the foundation of the elite family. (Ibid.: 43)

Such figures were members of the black elite. 'These people polished their domestic and social skills, while the majority – doing what they had to do in order to survive – neatly fulfilled racist prophecies of shiftlessness and immorality' (Barnes 1999: 94). But their way of life and their models of domesticity became the object of aspiration for a wider population. Thompson and Grace Samkange were widely reported on in the black press. Indeed,

> Some working-class migrants, particularly those existing wholly outside the formal economy, situationally used certain goods connected with 'modern' living to publicly signify their aspirations. These individuals often based their consumption habits on their observations of the African elite. (Burke 1996: 184–5)

In any case, for most of this emerging bourgeoisie, aspiration was a central part of class identity. Not marked by wealth, nor by descent from Rhodesia's pre-colonial rulers, 'as much as anything else, the African middle class in colonial Zimbabwe was held together by a unity of purpose: its members had interests, aspirations, and ideas that set them apart from other social classes, and they were conscious of these differences' (West 2002: 2). Unlike in some of the continent's colonies, 'cultural capital … consisted more of a Christian, bourgeois, achievement-oriented background than of lineal ties to the pre-colonial rulers' (ibid.: 59).

Attaining secure 'respectable' status was indeed often a battle, for early elites and later a broader aspirant middle class. For those hoping to achieve mobility through education, schooling could establish disorientating distance from one's own family, as Dangarembga describes in her novel *Nervous Conditions* (2004). Meanwhile, maintaining the necessary domestic ideal was a task that fell to women (West 2002). Despite the perceived moral taint of urban life, many endeavoured to spend as much time as possible with their town-dwelling husbands in order to discourage the latter from seeking other women, and to keep a conjugal house based around a resident married couple (see Barnes 1999). As in more elite cases, the idea of a strong marital partnership striving for family and domesticity, upheld crucially by women, was central to notions of respectability and middle-class masculinity. Further, spatial distinction was central. As Scarnecchia notes, 'the respectable classes actively sought to separate themselves from the lives of the poor and the uneducated in the townships' (1999: 161).

Reimagining Men 143

Education, marriage, domesticity, and distance from townships represented a recognisable package in Zimbabwean bourgeois consciousness (West 2002). While the prospect of realising this ideal was limited under settler rule, post-independence years saw 'unprecedented social mobility for the black majority, as legal barriers underpinning white privilege were removed, the state invested in education and the rapidly expanding public service was Africanised' (McGregor 2008: 469). Middle-class aspirations were easier to achieve. New markers of sophistication, in turn labelled through stereotypes such as *masalads* and *ma-nose brigades*,[1] did not render redundant existing and more widespread models of distinction.

Whereas some analysts have argued that elites are a more appropriate object of study in Africa than a middle class (see, for example, Chabal and Daloz 1999), notions of status, with greater emphasis on propriety and sophistication than on wealth or power, remain crucial to appreciating the self-understandings of many Zimbabweans – even following a decline in employment and educational opportunities from the 1990s and the consequent rise in material inequality. In recent years, better-off Zimbabweans have turned to places like the UK for livelihood possibilities (see McGregor 2008). An ever-greater number, with fewer resources but in some cases no lesser sense of their status, have made for South Africa, as we shall see below.

At Grootplaas, notions of respectability and sophistication offer a way to maintain a sense of distinction alongside undesirable farm work. But in its migrant form, such a model for conduct is more flexible. Those who adhere to it emphasise or de-emphasise different aspects according to their circumstances: education despite joblessness at home; a history of non-manual, or even skilled or artisanal, employment. Meanwhile, it is more tenuous, lacking the support of social circles and appropriate occupations. Further, those who understand themselves through this model differ in other ways. Some maintain strong connections with rural areas. Others grew up in Bulawayo or Harare and have more overtly urban backgrounds. But what I wish to explore in this chapter are enactments of middle-class respectability in the Grootplaas farm setting, which downplay such differences.

To appreciate the variety of forms such aspirations take at Grootplaas, it is worth beginning with non-seasonal workers. Benjamin, 35 years old in 2007, has a number of current or former teachers in his family – his mother and several siblings. Benjamin himself has A-levels. I accompanied him on a visit to his mother's home, in a rural area not far from Gwanda in Matabeleland South. As the youngest son, he had managed to

---

[1] The former referring to 'those eating salad, a foreign dish', the latter 'those speaking English in a very affected way, "through the nose"' (Veit-Wild 2009: 687; see also McGregor 2008).

provide for her, maintain the house that he will one day inherit, and keep a few cattle and goats. Despite this, his mother told me that she had hoped for more. Given Benjamin's obvious aptitude at school, she had wanted him to become a doctor. Benjamin has gained certificates in personal management, and has worked briefly as a teacher, but several failures and unfortunate incidents (notably his brother's death) led him to seek employment at Grootplaas. On the farm he is the storeman (in charge of allocating tools at the workshop) and sometime clerk, as well as a facilitator at the compound's adult literacy centre. Although these are fairly senior posts by farm standards, Benjamin often expresses disappointment at his lack of achievement (and is aware that his mother shares this sentiment). He complains that long working hours prevent him from reading, let alone studying, which might open up new opportunities for mobility. But he also often reflects on the considerable responsibility involved in his job, while enjoying the fact that his position means that he is usually left to his own devices. Meanwhile, like Michael, the personnel manager and another adult literacy centre facilitator, his authoritative, reputable role helping others to learn enables him to maintain a sense of respect. And the centre – 'the school' – is also a place where, each evening or over the weekend, he can discuss current affairs, the politics of language (English, SiNdebele, and ChiShona), or the excesses of other residents' drinking habits, with others who enjoy such debate. During my fieldwork, Benjamin lived alone and, apart from occasional assistance from female neighbours, he looked after himself. He subsequently found a partner and began the process of paying bridewealth. Importantly, he is known for being unusually unpredatory towards women; he does not drink (a family trait); and he invests significant amounts of time in intellectual self-improvement. Constrained by his living situation, he nevertheless strives to live by a model of temperance, restraint, and education.

Margaret's story, though different, presents obvious parallels. The 47-year-old daughter of a teacher and a taxi-owning businessman, she completed her A-levels, as well as diplomas in teaching, childcare, and social work. After 12 years as a teacher in Bulawayo and employment in Johannesburg at a crèche and in catering, she moved to Grootplaas to live with her husband, who had been employed as a security guard. She and her husband avoided being out in the compound, especially separately; whenever possible they would socialise with others who behaved likewise. Margaret would spend most of her time maintaining her husband's large, senior-worker house on the edge of the compound, and baking cakes which she would sell. Meanwhile, she maintained a sense of difference from other residents: she told me of a brief piece of research she had once conducted at the farm for a cousin at a non-profit organisation.[2]

---

[2] As noted in the previous chapter.

Lacking integration in the farm's work structure, and still well connected in Johannesburg, she often spoke of leaving. Eventually, her husband was employed as a security guard in Johannesburg, and they moved in with her daughter in Mayfair, west of the city centre. Where Benjamin had to seek a source of individuated, personal pride at Grootplaas itself, Margaret emphasised her professional background, a domestic and marital ideal, and continued connections in educated circles. These particular examples by no means exhaust the ways people act on their aspirations.

I began with the broad question: *what happens to class consciousness in conditions of displacement?* There is certainly little in the way of shared, workforce-wide class consciousness at Grootplaas – the basis for a clear dichotomy between 'capital' and 'labour'. This is because the workforce is full of attachments to past class positions and forms of class-based status – class consciousness in a rather different sense. Yet current occupation does matter, as personal histories and aspirations collide with present-day circumstances to produce class-based self-understandings in the farm context. Benjamin's storeman's and clerical employment distances him from team-based farm work, while his work as 'school' facilitator ensures him esteem as an educated man. Margaret's lack of employment, apart from the survey she had been employed to conduct and the research with which she assisted me, simultaneously allowed her to cultivate a self-understanding as a home maker while also denying her any strong attachment to the farm or its population.

Seasonal employees, to whom we now turn, are especially constrained by their current occupational circumstances. The adult literacy centre where Benjamin works does not offer them a way to underscore educational backgrounds or aspirations, since it is reserved for established residents. And, for most such workers, Grootplaas represents a stopover on the way to better employment. For them, preserving an existing sense of self is often more important than negotiating sustainable arrangements for farm life. In this respect, their models for life therefore tend to look more like Margaret's than Benjamin's. But, as the previous chapter showed, their limited resources make it particularly difficult to pursue a domestic ideal, as Margaret does. Meanwhile, seasonal workers face different modes of employment at the farm. Picking work involves a register of behaviour that presents a clear challenge to those with middle-class sensibilities.

Work and Masculinity in the Picking Process

The citrus orchards at Grootplaas – consisting mostly of Valencia oranges, with the remainder under grapefruit – appear as deep-green rectangular grids from the air. At ground level, arterial dirt roads, offering vehicles access, run between the blocks, composed of wall-like rows

Figure 5.2. Picking team.

of densely planted trees. Avenues between the rows are kept just wide enough for tractors and their trailers. At the peak of the harvest in 2007 there were 12 picking teams, each made up of 30 pickers (see Figure 5.2), of which ten teams were male.

Picking is a fast and aggressive affair. A pair of trailers is left in or at the end of an avenue of trees. Pickers carry large, square, plastic-covered canvas bags with shoulder straps (*zwigege*). Two team members, 'waiters', are designated to bring empty bags and carry full ones to the trailers. They sprint up and down the avenues, shouting 'Waiter!' repeatedly to announce their presence to pickers perched on branches or up ladders, who pass down their full bags. The latter, meanwhile, shout 'Waiter!' to call for a new bag. Others pick at ground level and may run their own bags to the trailers. While some pickers are difficult to see among the dense foliage, waiters remain in the open. Their effort to keep running and keep pickers supplied with bags is visible and crucial to the earning potential of the group. For pickers are paid per trailer, on a group-based piece rate. When the trailers are full, a tractor driver is called from the arterial road, his assistant hooks them up to the tractor, and they are hauled off. Meanwhile, the picking team is already seeking out the next trailers. Sometimes a pair is already waiting, placed by a diligent driver. Otherwise, the most eager pickers run to the road to call for one, and whistle a high, rhythmic monotone as it arrives to announce it and to regenerate the necessary intensity. A picker or two might climb into the trailer as it is hauled down the avenue, whistling and already cutting a few pieces of fruit. Others might have stopped to catch their breath, but by now the cycle and its pace have started again. When there is a backlog at the packshed there is sometimes a gap in the supply of trailers. Frustrating though this may be, it also offers an opportunity to rest, sleep if necessary, and eat some fruit. More often than not, though, breaks are kept to a minimum.

Shouting 'Waiter!' goes beyond simply coordinating bag conveyance. It provides a base level of noise that frames the rhythm and intensity of the picking process. This is widely supplemented by more explicit calls of encouragement by pickers and waiters: '*A ri thuwe!*', '*A ri dzheni!*' (TshiVenda: 'Let's go!'; 'Let's enter [the trees]!'); '*Famba, varume!*', '*Vari kudei?*' (ChiShona: 'Go, men!'; 'What do they want?').[3] Among the barrage of calls I heard were imperatives to keep working. There might be calls to work through lunch – this while the packshed and permanent workforce, including the picking supervisors and tractor drivers, take a break to eat. Entreaties to work fast and continuously sometimes take on a tone of moral appeal. On one occasion, two pickers were perched on branches in an orange tree. One told the other to pick faster, referring to the fact that they both had wives to look after.

A rough camaraderie characterises these picking teams of often young men. A picking day entails a long string of attempts to generate and maintain intensity, trailer after trailer, for up to ten hours. Work pace often slows in the afternoon as teams tire, but members are aware as the day ends how many trailers they have filled, and whether the number is sufficient to represent a decent wage. This self-imposed work pace and the generation of a work rhythm rely heavily on a particular mode of interaction in which dynamic productivity is connected with a performance of virility and physical power.

Narratives of misogyny and sexual promiscuity are established as means by which male workers can relate to one another and build a sense of collectivity through the work process. On the one hand, a physical, aggressive register of interaction helps workers to earn more: associating a sense of manliness with production promotes faster work and greater productivity in the team. But it also establishes common experience and self-understanding beyond work. For many men, agricultural work and farm living are unfamiliar experiences. Contrasting backgrounds can be obstacles to establishing relations. In this context, a set of easily learned, commonly repeated phrases during the work process, and modes of

---

[3] Both TshiVenda and ChiShona are used, but when one woman called for a waiter: '*Waiter papi? Waiter u gai?*' ('Where is the waiter?' first in ChiShona, then TshiVenda), Ezekiel, the supervisor, explained that she was Shona but trying to conform to the majority language by using TshiVenda.

behaviour suggesting a particular machismo, can act to bind strangers, albeit sometimes tenuously.

Similar notions of masculinity, generated through work processes in all-male environments, have been widely noted. In studies of Southern African music (see, for example, James 1999; Coplan 1987, 1991), performed masculinities produce imagined shared experiences. In an all-male part of the shopfloor in a northwest English truck factory, joking relationships emphasising male sexual prowess operate to produce conformity among workers: 'mediated through bravado and joking relations, a stereotypical image of self, which was assertive, independent, powerful and sexually insatiable was constructed and protected' (Collinson 1988: 191). A similarly sexualised male bravado obtains among black South African gold miners; indeed, 'sexual expletives and detailed accounts of sexual activity seem basic themes of underground conversation amongst miners all over the world' (Moodie 1983: 181). Male workforces in sectors as diverse as assembling trucks, underground mining, and picking fruit are built upon relationships of camaraderie rooted in sexualised masculinities. This, in turn, is perhaps due to the valorisation of bravery and physical strength. In the picking process at Grootplaas, such a performance ensures fast work and higher pay.

It also frames interactions between pickers and supervisors in the somewhat sheltered setting of the citrus orchards. One day, for example, a supervisor named Hardship was standing by the trailer with another, Ezekiel. As supervisors often did, they joked with and about certain pickers around them. Someone called to 'Superman' (a nickname), a picker nearby in a red cap with the word 'Marihuana' [sic] and a large green illustrative leaf emblazoned on the front. Superman is the twenty-something, younger brother of an established tractor driver at the farm, and was promoted to permanent worker status at the end of the season. The caller said to Superman that it was he who was responsible for a rape at Mopanekop, the neighbouring farm, recently. Supervisors and pickers laughed. Hardship and Ezekiel explained to me in some detail how two men had raped a woman there. The joke had been that Superman must have been responsible for this incident, because he had raped a woman at this farm eight years ago. He replied that that was in the past – he does not do that kind of thing anymore. Although distancing himself from this particular event, he did so with the same loud, joking tone as the other men, making light of the episode. Supervisors and pickers, in an all-male environment and sheltered by rows of citrus trees, established shared knowledge about Superman and his history, and contributed towards rendering the farm and its residents familiar to all involved. I never heard comparable conversations between seasonal workers outside the orchards. Unlike the New Houses in the compound, citrus avenues present themselves as a circumscribed space in which men can assume that only their picking team can hear them.

Developing good relations with supervisors is important, given their power beyond the work-time context. We saw in the previous chapter that *mapermanent* are at the centre of settled compound life. Senior workers, in addition, enjoy diverse forms of authority. The *khoro* – the community court over which the senior foreman, Marula, presides as a kind of *musanda* (headman) – is overwhelmingly made up of these picking-time supervisors (see next chapter). One is an influential United African Apostolic Church bishop (although he also drinks in the New Houses on occasions – unlike many self-consciously Christian compound residents, including those in his own congregation). Another runs the seasonal workers' football team and is involved in organising occasional concerts at the farm. How pickers engage with their supervisors during work time contributes to how notions of dominant male behaviour are constituted. Most supervisors encourage and engage in the performance I have described and so, for pickers, participating in it offers the chance to establish relationships with them. Conversely, the best-connected men are those who behave in this manner. On the day just described, supervisors joked with another picker, Tshigidi (literally, 'Gun'), Ezekiel telling me that he likes him a lot because he is funny. Tshigidi is his nickname and is intended to be understood in a sexual sense. He did, however, explain to me that he no longer wants to ask every woman to 'be his wife', as he had done in Zimbabwe, because he is worried about AIDS. Tshigidi's quiet aside to me, about his name and his reputation, underlines how the masculinity demonstrated through picking-team dynamics is a performance that may not entirely reflect individual members' own sentiments. Nevertheless, a coarse camaraderie connects certain pickers to influential supervisors. Tshigidi is an older man with many years of farm work experience in Zimbabwe. 'The farmers' (the whites at Grootplaas) joke with him that he was a war veteran, I was told, although he claims he was, on the contrary, working on a (white-owned) farm during the Zimbabwean liberation struggle. Tshigidi was upgraded to permanent status as a tractor driver towards the end of the harvest, benefiting both from his farm worker background and his easy relationship with supervisors. While very few seasonal workers are upgraded to permanent employment, establishing rapport with supervisors nonetheless renders their time at the farm much more liveable.

For the majority, work banter binds pickers together because it is not only a product of the work setting and its productive requirements, nor only a means to frame the wider farm experience: it also resonates outside work time.[4] Sex talk extends into compound life, albeit with more subtlety than in the apparently sheltered, all-male environment of the

---

[4] A methodological caveat: actual incidence of sexual activity is hard to research, and I, like others, have had to rely on how people talk about it. (See, for example, Campbell 2001.)

orchards. This subtlety is perhaps the result of the fact that, historically, 'the dominant African cultures in Zimbabwe placed a strong taboo upon the open discussion of sexual matters' (Epprecht 1998: 636). But it seems fair to suggest that some men behave at Grootplaas in ways they would not at home.

The widespread celebration of highly sexualised – even predatory – behaviour in male workforces appears to suggest that such environments tend to elicit these forms of interaction. Some scholars posit even more direct connections between work and masculinity. Campbell argues, of South African miners, that 'male identities serve as a key coping mechanism for dealing with high risk working conditions, through encouraging men to be brave and fearless in the interests of supporting their families' (2001: 276). Paradoxically, these identities promote risky promiscuity and lack of concern for health consequences. I do not wish to make such an argument about pickers at Grootplaas. Masculinity *is* constituted through the work process, but also reflects a *wider context* at the farm.

It has been noted for other places that economically motivated migration, while uncomfortable, can come to offer other attractions.[5] Visitors to Grootplaas would comment to me about the available women, suggesting opportunity or danger.[6] Women in relationships with men would complain about the large number of unattached women who come to the farm, suggesting even that they come to stir up trouble (see previous chapter, also Rutherford 2001). And in one evening's conversation I was told, perhaps with exaggeration, that men might share their girlfriends with 10 or 15 other men; of the large number of people at the farm with 'the epidemic';[7] and how sexual relations are related to fights in the compound. Seasonally employed men, away from home and often far from any relatives, are able to take advantage of their residence in a transient population, where women are also away from home and possibilities for recreational sex abound. Yet men often arrive with no resources. Paid according to a piece rate, male pickers tended to earn in the mid-R800s during the period of fieldwork, less than female packshed workers who are paid the hourly minimum wage.[8] Along with their sparse living conditions, shared accommodation, and extremely uncertain futures, a lack of material resources limits seasonal workers' chances of establishing domestic (as opposed to merely sexual) arrangements with women. For seasonal workers, the political economy of sexual and domestic relationships is one in which sexual encounters tend to be fleeting and lack

---

[5] See Ventura 2006 for Filipinos in a Japanese slum; Lindquist 2009 for Indonesia.
[6] Of sexually transmitted infection.
[7] HIV/AIDS.
[8] Minimum wage R989 during fieldwork. Of seasonal employees, packshed workers earned around R1,000 because of overtime. Female pickers could earn as little as R300, because they were paid on the same piece rate as men but picked more slowly.

Reimagining Men 151

wider domestic obligations in sharp contrast to those of *mapermanent*, encountered in the previous chapter.

What, then, are we to make of the dominant form of camaraderie in male picking work? It reflects wider lifestyles, and mobilises these in generating shared experiences and bonds between workers. Indeed, it comprises a statement about what it means to be male at the farm. And, as I show in the next section, it is against such a performative statement that a minority of men define themselves. For those who do participate, picking-team work banter facilitates cooperation; for those who opt out, it does the opposite. Anthropologists of performance in South African migrant labour settings have illuminated how musical performance can be central to imagining social groupings and conceptions of class.[9] The picking process is no different in this regard, as it aligns men with varied backgrounds according to two divergent masculinities, themselves reflective of models of class reimagined in the farm context.

'They have Offices in the Trees': Models of Masculinity, Ethnicity, and Class

Alex and Vusa were young men in their late teens who had come to Grootplaas together from Bulawayo in 2007; Simon, 30, had come from Harare and met the other two at the farm. The three lived together in the New Houses. Alex had just completed his A-levels and wanted to save money for a computer to start a DVD rental business. Vusa and Simon were also well educated. Like many others, all three found living on a farm a completely alien, and often alienating, experience: Vusa wondered whether he should take photos to show people at home, and write a book about his experience. In the evenings they would cook outside their room on a fire, and sit inside on the floor along the walls chatting, mixing languages, and peppering their English with Americanisms to display a 'hip' urban sophistication (see Veit-Wild 2009). Sometimes conversation would turn to working at the farm. They would imitate Venda phrases and comment on behaviours they found aggressive among 'people here', such as greeting by exclaiming 'Yah!' One evening I was in their room where I had been interviewing their 23-year-old friend, Tendai, who had also reached A-level. A broader conversation developed. He said he had read management books and knows that managers ought not to manhandle workers. But they get pushed and kicked, he said, by other pickers. Most foremen here are from a particular village near the border. Pickers from these border areas therefore feel that they are superior and can boss others around. 'They have offices in the trees,' he said. But such people are not well educated, he continued. They treat the

---
[9] See, for example, Erlmann 1992, James n.d.

whites on the farm with undue deference, like gods. The implication was that Tendai, Alex, Vusa, and Simon, as educated people who were the equal of whites, would not need to display such toadying deference. Alex offered in response how *he*, by contrast, had merely thanked the white production manager when the latter complimented his picking. Others had asked him, in hushed tones: 'What did he say?' Alex told me that 'the foremen [supervisors] expect we'll treat them like they treat the white guys', and do whatever they say, even if it is wrong. 'But we say, "No, we won't do it like that."' Vusa proceeded to assert that the uneducated at Grootplaas could not argue, but instead just got increasingly emotional. Tendai agreed. Earlier, I had asked Tendai for what reasons he might consider leaving the farm. He answered by saying he would leave to find a more skilled, more rewarding job. Reflecting how, for them, picking-team dynamics were both alienating and also offered stereotypes against which to maintain a sense of difference, Alex and Vusa imitated the picking-work call – 'Waiter! Waiter!' – and laughed.

For such young men, living in groups in the New Houses, but with middle-class models of behaviour, the work process constitutes an object from which they can distance themselves. Picking presents them with a concentrated, stark form of a stereotype of aggressive, crude, Venda men. It exemplifies for them what Venda men are like more generally, a formula against which to define themselves in the farm setting. This foregrounds the importance of their own education. While Tendai cited management texts as a vantage point from which to cast Venda men as inferior, Vusa spoke of writing about his experiences to communicate to sophisticated friends at home what 'people here' are like. A particular version of respectability emerges through their comments. Unlike the rough behaviour of the majority of men, they characterised themselves through education and etiquette. They spoke of their time at the farm as a difficult period; one which would by implication stand in contrast to their urban, cultivated lives. This was underlined by stories of home, which involved driving cars (remember that they are barely beyond school-leaving age) and, in Alex's case, these accounts were illustrated by a photo album of his well-dressed family. This album was Alex's continued connection to the kind of family ideal that West (2002) describes.

They are not alone in framing class models through ethnic difference. George is a Shona: a Rastafarian musician with A-levels from a multiracial (that is, formerly 'white') school. He sought work as a picker at Grootplaas in 2007. Two years before, his handicraft business had been destroyed during the *Murambatsvina* 'slum clearance', and he eventually decided he had no other option but to jump the border. After the 2008 elections, George returned to Grootplaas. His status as an MDC activist now gave him added cause to leave Zimbabwe, given the wave of state-sponsored violence against those who voted for the opposition.

His reasons for being at the farm – the collapse of his livelihood and his fears for his personal safety – illustrate the simultaneously economic and political nature of displacement.

George and his housemates – friends from home (one with primary education only, the other up to the year before O-level) – told me their opinions about Venda people. They are vulgar, violent, and their women engage in prostitution. The very sexual relations that are so often the stuff of picking dynamics are here cast as the marks of an ethnic other. George commented, in a conversation with the other two, that the difference can be explained as a result of educational hierarchy: Shona are the best educated, followed by Ndebele, with Venda as a poor third since they always left school to work on the farms. This followed the trio's observation that they stayed inside their house in the evenings because they feared unprovoked violence by Venda. The problem, they said, was 'tribalism' (a word I often heard used). But education here once again operates as a marker of ethnic difference.

In a discussion of similar kinds of 'traditionalist' versus 'respectable' dichotomies, James (1990) highlights how ethnic stereotypes, in an area further south, mobilised different historical events and processes, making them grist to the mill of present-day concerns. At Grootplaas, education and notions of class are intertwined with ethnicity in the stereotypes held by farm residents. Venda from the border areas tend to have lower levels of education than people from areas further north in Zimbabwe.[10] Marginalised within Zimbabwe, their work prospects seem to lie on the border farms, and they therefore lack incentives to push for other opportunities. I was told that many farm residents from the far south of Zimbabwe leave school early and cross the border to get a job. Girls in their early teens cross to look after the children of family friends or relatives who work on the farms, thus earning a bit of cash for their own parents. Meanwhile, an established history of work on South African farms – work of extremely low status and symbolic of a past racial order – contributes to the depiction of border-dwelling Venda as overly deferential towards whites and, in a sense, as colonial relics themselves. More generally, these stereotypes associate Venda with, and revolve around, interpretations of the masculinity generated through the picking work process. The typecasting does not run only one way. Non-Venda from further north are portrayed as taking employment opportunities on the farm previously enjoyed exclusively by southern Venda. For some, resentment at the flood of Shona work seekers is augmented by the perception that it was Shona

---

[10] For example, of the sample of 79 permanent workers surveyed in 2008, 17 out of 23 Shona respondents and 10 out of 15 Ndebele respondents had been to secondary school (that is, had received eight years or more of education). Only four of 24 Venda respondents had been to secondary school.

who ruined Zimbabwe in the first place. This Shona dominance, like former class relations, is now reversed on the farm.

For some of those who consider themselves more sophisticated, complaint in the shelter of their rooms is not enough. While the examples I have cited so far involve men living at Grootplaas without women, a minority come with spouses. Chipo, 31, was a businessman who arrived at the farm with his wife, Siyanda. Previously he had come to the farm to sell cigarettes, and was now continuing to supplement his wage with cigarette sales (see Chapter 7). Even so, he was keen to assert a sense of difference, which he highlighted by dressing in a grey suit and brown leather shoes on his final day at Grootplaas, and he was quick to point to others' lack of education. Although he described himself as a drinker, he would often be in his room in the New Houses with his wife, distancing himself from the stereotype of the predatory man who consumes vast quantities of beer and dances into the night. Like Margaret, mentioned earlier, he sheltered his marriage from compound life, and described the close companionship with his wife as based on deep mutual understanding. Assertions of ethnic difference are perhaps foremost among men living in groups; but for those who are a little older, and accompanied by spouses, the defence of a particular conception of domesticity becomes more important. One of his closest companions was Jameson, a man who placed a similar emphasis on respectability, and on his own gendered role. I now describe Jameson's case in somewhat more detail, since he most clearly exemplifies the predicament of trying to preserve a marital ideal.

Jameson was 40 in 2007. He had come to Grootplaas with his wife, Jenny, from Zvishavane, where they had lived with their two children along with Jameson's sister's son and his father's younger brother's daughter. He had not only succeeded in passing his A-levels, but also completed a Bachelors degree and a postgraduate diploma, as well as a certificate in entrepreneurial development which he undertook in Nairobi. He had worked as a teacher, as well as in a post office and for the Zimbabwe Election Support Network and the National Constitutional Assembly. Jenny had been to a mission-run secondary boarding school, where she passed her O-levels. Both speak ChiShona and English. Like many others, they left to come to South Africa because the Zimbabwean dollar had deteriorated to the point where teaching did not provide enough for Jameson to support his family. There was also perhaps another side to their migration, which neither articulated explicitly: Jameson's employment by the National Constitutional Assembly would have marked him out as a political dissident, leaving him potentially vulnerable. They had travelled to Beitbridge, where they joined others, to go to a farm on the Zimbabwean side of the border. There, they were told of job opportunities on the South African side. Jenny was employed as a cotton picker on

Figure 5.3. The object of avoidance: dancing and drinking at the New Houses.

one of the border farms; Jameson heard that Grootplaas was recruiting citrus pickers.

Jameson and Jenny displayed a self-consciously Christian sense of propriety, and attempted to maintain the conditions in which their established notions of proper conduct could be enacted. Luckily for them, Jameson's brief stint in the packshed (see below) had secured them a room outside the New Houses. They opted out of much of the dominant social life of the compound, avoiding drinking (he does not drink beer, only the occasional glass of wine), dancing, and promiscuous behaviour (see Figure 5.3). They and others with similar views would keep their distance from the New Houses at night when possible, and Jameson told me that he would go to church on Saturday nights when he got the chance – in Jameson's case, to the Armed for Harvest[11] services held in the compound hall. Jameson and Jenny's effort to maintain a class-specific lifestyle by asserting a sharp withdrawal from compound dynamics mirrors strategies of spatial distinction in Zimbabwean historical imaginations of class (Scarnecchia 1999; West 2002).

---

[11] The name of a church.

The two couples – Jameson and Jenny, and Chipo and Siyanda – despite their short stay at the farm, worked to establish a degree of domesticity in their rooms. Jameson and Jenny had rigged a piece of net cloth to separate their sleeping area from the rest of the room, and they bought an electric camping stove to cook inside. Chipo and his wife had done something similar. And when I visited, they would either invite me inside the room (in the evening; both couples) or offer me a soft drink and biscuits outside on chairs or stools (only Jameson and Jenny had the facilities for this). Their use of space (inviting me, and other guests I brought, inside the room), and in Jameson's and Jenny's case their style of refreshments, demonstrated through their house-keeping and hospitality both 'domesticity' and 'modern respectability' (see Rutherford 2001: 120).

Like Margaret, Chipo, and their spouses, and emphasising their difference from many others, Jameson and Jenny avoided any extended separation. They would spend every evening in their room together. Jameson's presentation of his relationship with Jenny tells us something about how he understands himself. Jameson would generally refer to Jenny simply as 'my wife', rather than by name. In doing this, he foregrounded the formal status of their relationship in a way I seldom heard, except in the case of Margaret and her husband. Jameson would present his marriage as a partnership of joint, coordinated effort and decision, and this inflected his and Jenny's discussions about how to proceed after their stint at the farm (they intended to work in Pretoria, Jameson possibly going ahead to find accommodation). Crucially, this was a marital ideal under threat: Jameson felt that, now that he knew what went on at the farms, he could never leave Jenny unaccompanied here.

But it was in picking work that Jameson had most sharply to contrast himself with others. His pleated trousers, tucked-in, long-sleeved shirt, and gold-rimmed glasses set him apart from other pickers. He soon established a reputation in the team and throughout the farm. Everyone knew him as Magogoros – 'Glasses' (literally, 'Goggles') – because he was the only person in the compound who wore glasses all the time. His quiet voice, reserved manner, and occasional correction of workmates' factual assertions compounded a professorial stereotype. Within a couple of weeks, he was promoted to a clerical job in the packshed after the personnel manager, Michael, noticed him. Jenny was also promoted to a job as a fruit packer in the packshed. However, Jameson was soon demoted and sent back to picking because of personal differences with Michael, rumoured by some (including Jameson himself) to be a consequence of Michael viewing Jameson's education as threatening his own educated status and his job.

Back in the orchards, he told his team mates to stop addressing him in the same crude tone as they did each other. He was particularly disgusted by references to one's 'mother's birth canal', as he expressed it. His attempts to explain this in his own words are evidence of his profound feelings of unease with the tone that flavours the constant stream of banter accompanying work. On the same day, he reported, one man had commented he would not return home in the two-week break between the grapefruit harvest and the start of the oranges; another asked him: 'Why? Did you eat your mother with your penis and are worried about it?' Jameson was upset that men around him felt they could speak thus to an older man, although there were older pickers who participated in the banter.

In this way, Jameson, in effect, opted out of the work process, and he and one other man who took offence were posted to clearing the trees of oranges the team had missed, thus working at a distance from the rest of the group. He later also worked loading fallen oranges onto trailers, with others who had been 'chosen' by their supervisors. George was among them. Conversation in this work team was dominated by discussions of the educational hierarchies between ethnic groups, which George had described to me elsewhere. They also addressed Zimbabwean politics, more so than any other seasonal workers with whom I spoke. This is perhaps unsurprising, given their backgrounds. We have seen that Jameson had worked for the opposition National Constitutional Assembly, as well as for an organisation that promoted electoral freedom. George had lost his shop in *Murambatsvina* and had later been a vocal supporter of the MDC opposition. But discussions of politics were equally a means of asserting difference from other workers. Through these conversations, they shared a sense of educated understanding of the big picture of Zimbabwean politics. The tone was one of detached analysis: was Morgan Tsvangirai, MDC leader, sufficiently educated to run the country, or should the role fall to rival Arthur Mutambara, on account of his PhD?

Picking teams operate according to a particular masculinity. But, clearing reject oranges, Jameson and George asserted and performed their own, contrasting self-understandings. This was in turn paralleled outside worktime. Perhaps unsurprisingly, one man with whom Jameson did enjoy talking in the compound was Benjamin, the storeman described earlier.

Connell (2005) contends that different masculinities must be understood in relation to one another, ordered through unequal power relations. Those described in this section present and understand themselves according to a notion of masculinity that might have asserted its superiority in other settings, but is marginalised by the more dominant version evident in the picking process. They understand themselves through culturally dominant ideals of having succeeded – as having attained the credentials and manners of the Zimbabwean middle class (see West 2002) – while now lacking the material and occupational bases of this class position, as well as social networks and other means of power associated with this background. Modes of self-presentation, which may have contributed to reproducing relatively privileged positions at home, do Jameson and others little service in a farm workforce. At Grootplaas, they find themselves faced with a worktime register of interaction between men that, making explicit how these other men spend their time outside work, disrupts their ideals of appropriate male behaviour.

Conclusion

Zimbabwean male seasonal pickers at Grootplaas enact divergent masculinities along class lines, sometimes also framed in ethnic terms. This reveals that models of middle-class respectability powerfully shape how residents of Grootplaas distinguish among one another. Departing from a historical perspective of Zimbabwean middle-class consciousness, and exploring how a bourgeois model is mobilised by some in the farm context, I analysed the picking work process as a performance of opposing ways of being male. For many male pickers at Grootplaas, a particular sexualised register of camaraderie establishes bonds at work that reflect realities in the compound. It establishes shared experience amidst diversity. And it builds good relations with powerful supervisors. The interactions involved also appear to establish the basis of effective teamwork. But some men, understanding themselves through a notion of respectability, are alienated both by the style of the work process and by the compound life that it reflects. Important models of class difference are at work among residents, who polarise according to them. These models in turn both reflect and shape experiences of Zimbabwean displacement.

The next chapter turns to the most senior members of the black workforce. Building on this chapter's analysis, it examines how class-based aspirations, forged in the context of Zimbabwean displacement, shape the meanings of contrasting models of agricultural labour relations: paternalism and corporate-style managerialism.

6 'Management' or 'Paternalism'?
Race and Registers of Labour Hierarchy

Introduction

The sharp spatial separation of white farmhouses and black compounds is a stark reminder of class and racial inequality at Grootplaas, suggesting a familiar story of outmoded estates as remote, totalising worlds. Farmers appear to fit popular conceptions, with their uniform of two-tone khaki shirts, rugby shorts, and occasionally hand guns – emblems of a vilified settler history. But white agriculture has seen important changes. Farmers have adapted to new times. Their iconic khaki shirts often now display the logos of export agents. While commercial farms remain divided between white farmers and black labourers, reframing labour arrangements in limited ways can be useful for farmers. Historically, farmers saw themselves as paternal figures, safeguarding the lives of their black 'people'. (Such assumptions sounded benign, but the presumptions about a wholly owned workforce also served to justify, or hide, abuse.) This is the view that underlies farmers' musings on racial difference in their 'pioneer stories' (Chapter 2). Farmers are now keen to assert that their enterprises are businesses like any other, that laws of supply and demand and budgeting concerns – rather than labour relations – are paramount. Squeezed by a liberal buyers' market on the one hand and post-apartheid minimum wage and housing legislation on the other, they are increasingly unable or unwilling to continue promoting themselves as their workers' fatherly protectors.

The result is, broadly, a shift from 'paternalism' towards an emphasis on a style of 'management' that appears neo-liberal. Responsibilities that are not explicitly guaranteed by contracts or laws are abrogated. Existing scholarship has been concerned to define paternalism itself, in order to specify what has changed. Du Toit (1993) sees it as a hegemonic discourse in which farms are presented as being united in the face of outside threat rather than divided by class contradiction. 'Paternalism' here appears as a coherent framework of shared understandings among farmers and farm workers. Its diverse relationships become reified as a system of sorts, that may give ground to, be replaced by, or articulate with new arrangements, such as 'neo-liberalism' (see, for example, Addison 2006).

It is difficult, however, to distinguish one discursive regime from another in practice. Attempts to define paternalism have oversimplified an amorphous and ever-changing array of work arrangements. They have suggested stable arrangements, whereas farming areas have long been characterised by the fragmenting and atomisation of black families in a mobile population (see, for example, Waldman 1996). At Grootplaas, as I have shown in previous chapters, permanent workers establish a degree of homeliness, adapting accommodation and maintaining domestic arrangements with kin, visiting spouses, or partners in relationships that have permanence only at the farm. These possibilities are denied to seasonal workers by their transience and overcrowded, shared housing. This suggests that whatever paternalism does exist is restricted to the permanent workforce. However, picking supervisors establish themselves as important pastoral figures with wide-ranging responsibilities beyond work, offering quicker access to documentation, use of vegetable gardens, and safe places to keep earnings. A regional history of workforce fragmentation on the farms coexists with continuing vertical dependencies among workers – mediated paternalism – belying any simple shift from paternalism to a more atomised style of managerialism.

Moreover, recent changes in agriculture do not have only disintegrative effects. They also create new bases for hierarchies and arrangements between workers. Much of the shift towards corporatisation on farms is driven by links between farmers and supermarkets, but these relationships nevertheless potentially create new vertical dependencies within workforces. Farmers become more integrated into relations with 'capital upstream and downstream of farming' (Bernstein 2007: 40). Reflecting this, supermarkets now establish development projects on farms for workers. These enable supermarkets to sell 'ethical' produce while tying farmers into supply chains. In farm workforces themselves, key workers act as gatekeepers, controlling access to the resulting facilities – crèches, literacy centres, games rooms, and football equipment. At a farm like Grootplaas, such gatekeepers strive thereby to gain authority in the time beyond working hours.

A useful way of understanding the complexities of change on white farms is to begin with the relationships and dependencies among workers and other compound residents. Doing so avoids the risk of presenting change merely as top-down, driven by farmers or outside forces, and losing sight of farm workers' diverse circumstances, agendas, and projects. One study of paternalism that takes this kind of bottom-up view is Rutherford's account of 'domestic government' on Zimbabwean farms, which sensitively illuminates different workers' arrangements and plans (2001: 14), all broadly embraced within the paternalist model. But, unlike Rutherford's Zimbabwean case, in contemporary South Africa paternalism is seen as a distinct regime opposed to 'management'.

As farmers have been pressurised to corporatise, 'paternalism' and 'management' appear as two competing, and incompatible, models of agriculture.

For paternalism and management are not only analysts' models. This chapter explores how they are interpretive tools for farm dwellers themselves. At a time of uncertainty and contestation within white agriculture, management *and* paternalism receive support from different quarters at different times. Contrasting visions and interpretations of white agriculture are emphasised and de-emphasised strategically by particular farmers and employees. A history of regional farming leaves paternalism and management standing opposed as clusters of ideas not only about the scope of vertical obligations and dependencies, but also about how hierarchy, race, and class are imagined more broadly. Paternalism is a recognisable *ideal*, drawing explicitly on racialised, generalised authority and responsibility to subordinates. It is constantly being reinvented to justify decisions, actions, and arrangements (James 2007: 232). Managerialism – emphasising bounded work responsibilities and a deracialised management/labour distinction – serves equally justificatory purposes, both within and outside workforces. Its use as a justificatory measure is impelled by pressure to change, which comes from various external sources: government, buying agents, and supermarkets. Yet these paradigms, even when they are mobilised only strategically, suggest forms of proper authority that come with ethical entailments. It is by means of these models, and the contradictions between them, that senior figures on the farm may be held to account.

Workforces are more individuated than they initially appear to the observer or the reader of existing depictions. Appreciating this reveals that, rather than being merely top-down, hierarchies are most intensely shaped by particular senior black employees, whose status is most at stake. It is for senior black employees, especially, that paternalism and managerialism serve as models through which the meanings of race, class, and status are played out. Appreciating their roles in shaping labour hierarchies reveals how the broad shift from paternalism to managerialism in white South African agriculture is refracted and constituted in the dynamics of workforces themselves.

This chapter is organised around a rivalry between the two most senior black workers at Grootplaas: South African Marula, the foreman, and Zimbabwean Michael, the personnel manager. Their enmity reflects their different positions: top supervisor out in the orchards, and office-based administrator. Each attempts to undermine the other through belittling comments: Michael is presented as unreasonable in his dealings with subordinates; Marula as stupid and uneducated. Underlying this is the threat each presents to the other's status. Marula worked his way slowly up to the position of foreman, loosely modelled on that of village headman. This position, formerly known as 'bossboy', is central to the paternalist vision of agriculture in South Africa and Zimbabwe, and equally iconic in literary representations of white farming:[1] the intermediary father figure with broad authority and responsibility over workers. His very name, with its nickname-like connotations (Marula is an alcoholic drink – see Introduction, p. 2, footnote 2), belongs to a register that speaks of a life on white farms. Marula's status is directly challenged by Michael, who casts his professional success at Grootplaas in a globalised, dehistoricised managerial idiom, avoiding specifics of farm work. With his non-farm background, his desk job, and his formal qualifications, Michael desires to be regarded as not 'just' a farm worker.

Although their perspectives are expressions of individual personality, they draw on available discursive resources shaped by wider circumstances. Their conflict reflects, in one workforce, a struggle between contrasting ways of seeing white agriculture in South Africa. The positions 'personnel manager' and 'foreman', drawing on competing languages of hierarchy, cannot be easily or simply ranked vis-à-vis each other. Expressed through the wrong register, each man's occupation looks distinctly inferior. As I was told by one worker, Marula is the man on the ground, the man who deals with people. In contrast, Michael is a distant figure of command – like a white man, the boss, and a superior on that score. Michael asks for ten more workers; Marula finds them. But a worker from a neighbouring farm offered a contrasting interpretation when he unwittingly provoked Michael's ire by calling him 'clerk of Grootplaas'. This contestation produced heated debates between these two senior workers and other residents, in which I became implicated. For, although I occupied my own room in the compound, I lived as a satellite of Michael's household. Michael's responsibility for my welfare became a way for him to set himself apart from others in the compound, as the guardian of an academic researcher. But I would also while away evenings and weekend afternoons at Marula's house, or at the premises of one of the compound's many beer sellers. Residents often voiced their views about one or the other figure of authority, judging them according to competing notions of hierarchy and responsibility.

The chapter explores how these apparently 'old' and 'new' frameworks – paternalism and management – are both employed concurrently in this case of large-scale transformation. The enquiry draws on the kind of interactionist perspective exemplified by Epstein's (1958) work on copper mining in Zambia. Epstein challenged 1950s teleological depictions of urbanisation and modernisation through close attention to the roles and perspectives of intercalary figures – senior members of the workforce.

---

[1] These include J. M. Coetzee's *In The Heart of the Country* (1976) and Doris Lessing's *The Grass Is Singing* (1950).

He showed forms of authority with different sources, conflicting and pertaining to different spheres of life. A gradual shift from 'tribal' to class-based authority structures on the Copperbelt between the 1930s and 1950s involved competition between different bases of hierarchy that, crucially, were *all products of the mine setting*. As on the Copperbelt, those with intercalary roles on white-owned farms like Grootplaas are the senior black employees. They represent the interface between the black workforce and white bosses and senior managers, who perceive one another largely through the narrow lens of employment relations. Their interface is one of both work and racial categories, deeply intertwined – bosses and labour, white and black. This class/racial division inflects how 'paternalist' and 'managerial' ideals are understood as they are evoked and enacted by both employers and intercalary employees. At Grootplaas, a situational perspective reveals farmers even on one farm combining registers differently depending on contingent priorities: projecting an image of themselves or the farm to the outside world, or governing workers. And it reveals how intercalary workers – like Michael and Marula – shape the everyday meanings of labour hierarchies to maintain their own status.

This approach illuminates both the (global) effects of a shift towards corporate norms and the (local) effects of Zimbabwean displacement on the border farms. Conflicts over models of labour hierarchy mobilise diverse resources, including those from top-down interventions like supermarket projects in the compound. These conflicts play into further struggles over class-based status, explored in the previous chapter. Competition between the status that comes from schooling and that emanating from workforce seniority plays directly into the competition between managerial and more established idioms of authority. As we have already seen, large-scale displacement from Zimbabwe has meant that the border farms' black workforces are extremely diverse in terms of class-based aspiration. Some new arrivals are unwilling to accept the racialised, generalised forms of authority and responsibility – 'paternalism' – seen as normal and even laudable by many established residents. For such workers, evoking a corporate-style rhetoric – which itself stands as a mark of wider experience and schooling – can offer a resource for reinterpreting labour relations. The experiences of Michael and Marula, and the ways they are interpreted, both reflect a response to Zimbabwean displacement. Locally powerful norms of seniority on a relatively secluded farm confront the wider reach of educational and occupational status. The following sections describe Marula's and Michael's work roles, before turning to a consideration of their authority beyond work. The chapter then interprets these competing registers of hierarchy in terms of the highly situational priorities of Marula and Michael – and also of different white farmers.

Marula: One Man and His Bakkie

Each day, Marula ranges across the farm, visiting various workers and work teams in his *bakkie*. A man in his late forties, his voice husky, one eye milky from cataracts, he communicates a gravitas not to be doubted in the Grootplaas setting – despite the fact that his pink bomber jacket, worn baseball cap, and old T-shirt lack the sartorial elegance of some younger workers. His presence is vastly augmented by his *bakkie*, handed down to him via farmers and then white managers.

On a fairly typical work day in February 2007, Marula began his afternoon by driving to some orchards. Six permanent workers, aided by the dragging power of a tractor, were cutting down a row of pine trees to allow the adjacent orchard greater access to sunlight. Marula watched for a while. André, the white production manager, stopped by. When Marula was warned that falling trees might hit his *bakkie*, he drove to where another group was clearing debris, in an orchard where mangoes are grown for consumption on the farm, and gave workers a lift back to the compound. On his way to the eastern part of the farm, Marula drove a group of women to the army station: they sought access to the Limpopo for a fishing expedition. He turned off into orchards to visit the old women who were posted at intervals to look out for baboons stealing fruit. In the compound Marula bought three cigarettes and, patrolling around, shared these with groups of male workers. As the time approached five o'clock, Marula made one last trip to the orchards, dropped tools at the workshop, and called it a day.

Marula moves around to supervise different jobs, but in many cases there is already a supervisor on duty and tasks are fulfilled adequately without him. At some level it is difficult to see what he does. But his job consists in his place in the labour hierarchy, at the interface between white and black. Marula, one of the farm's few South Africans, speaks Afrikaans, and so can discuss matters in whites' mother tongue without the Zimbabwean majority understanding. But most of the time, Marula does not work through face-to-face interactions with whites. His very responsibility consists in working on his own initiative; his effect is to *extend* the labour hierarchy – and, by implication, the farmers' gaze – over a wide area, as he moves about the farm. By doing favours along the way – giving a lift, sharing a cigarette – he maintains a position that is both superior and sympathetic, a fatherly figure in the workforce. Meanwhile, as the radio in his *bakkie* crackles constantly with communication between the farm's vehicles and the office, he brings with him the institutional weight of the white employers. Other farms along the border have similar central figures, whose job it is to have a broad understanding of the whole estate: the foreman of a neighbouring farm, for example, made a point of telling me, when I visited, that the reason he already knew who I was, was because it was his 'business to know'.

Marula's role as lynchpin, necessary because of a rigid black/white separation, is intrinsic to the racial micropolitics of farms. Whites on the farm rarely enter the compound; it is Marula who usually gathers workers, including for emergency jobs in the evenings, at night, or over weekends. This divide is characterised by constrained forms of communication. The farmers address most black residents in a pidgin called Tatelapa.[2] This is the agricultural equivalent of Fanakalo, a hybrid language developed on the mines largely as a means for whites to direct black subordinates and lacking much range of expression. But even when issuing commands, the farmers speak to few workers on a regular basis. Generally, supervisory employees direct the wider labour force. These are Marula himself; two more junior, specialised foremen; and André, the white production manager. Willem, the white farmer, who spends much of his time around the offices, represents a distant authority. He is more a last resort for complaints or appeals for leave than a direct overseer of labour.

It is between Marula and the white manager, André, that the white/black divide is most blurred. André speaks TshiVenda and gossips with workers out in the orchards. His and Marula's roles often appear similar: each drives around in his *bakkie*, supervising a range of tasks; each negotiates his authority through a mixture of command and easy informality. Yet, within the world of farming in the region, they are vastly different. André's involvement as a white employee places him as a relatively poor man in the farmers' social circles – in comparison to local employers who own aeroplanes to go with their farms. Not unreasonably, permanent workers joke that he is too poor to buy his own farm. Yet André is unambiguously one of the local white community. Marula, meanwhile, has very different expectations. His occupation as foreman places him in a patriarchal role that extends into the compound. This is the culmination of a life climbing upwards through the ranks of the workforce, in a highly racialised setting where he has risen as far as he can. Despite the day-to-day similarity between Marula's and André's roles, the ways they are seen by everyone on the farms – and the ways they see themselves – reflect and reinforce notions of racial division.

Marula's position in the orchards – in relation to black workers and to André – emerges most clearly during the harvest. Picking work, mostly in 30-man gangs, is fast and aggressive; as the previous chapter showed, it often forges strong bonds between recruits. Supervisors participate in the crude, highly energised banter. But Marula stands above all of this, as he patrols the orchards in his *bakkie* at his own discretion, jokes both with André and the supervisors who work under him, and shares a newspaper with them. He drives the work pace not by continual shouting, as the supervisors do, but by highlighting the potential vulnerability of seasonal

---

[2] Also sometimes called ChiLapalapa, using the ChiShona/TshiVenda grammatical form.

Figure 6.1. Marula and André gather pickers for a meeting.

pickers: if they do not follow orders – correct picking technique, keeping to the correct route between compound and orchards – they can go back to Zimbabwe or find employment elsewhere. Once again, he evokes the larger institutional power of the farm.

Marula often announces a briefing, generally responding to white management's complaints about pickers (see Figure 6.1). Sometimes André stands a short distance from him, backing the weight of his words but also introducing ambiguity as to whose words they really are. Nevertheless, Marula's speeches demonstrate his own authority and that of his supervisors. This is emphasised by the form such addresses take. He often drives up in his *bakkie*. His supervisors are on the back, sitting on the sides or standing and looking down at the waiting 360-strong picking force, as though riding a chariot. To speak, he disembarks and stands by or on it, the supervisors often in a group nearby, apart from the pickers. When talks are given at the end of the working day, the *bakkie* pulls away in the direction of the compound, kicking up dust in the wake of which the column of pickers embarks on the long walk back.

Marula's flexible yet racially inflected authority must be understood in the wider context of farm labour in general at Grootplaas. As noted in Chapter 4, the farm manages with a relatively small permanent workforce by keeping workers flexible. A worker's level of responsibility is usually more important than his job description. Drivers are higher up than run-of-the-mill labourers. And the driver who operates not only tractors or forklifts but also lorries and other vehicles is superior again. Such responsibility is reflected in pay. Marula's role as foreman is as diverse and flexible as the workers he supervises. It places him at the centre of the labour process, and he constantly works with André, permanent workers, and, during the harvest, seasonal pickers, holding the labour hierarchy together.

Michael: The Power of Paper

Michael's position contrasts starkly with Marula's. Each work day just before seven o'clock in the morning, Michael cycles into the office at the farm's workshop. He wears a collared shirt, often white, tucked into a belted pair of dark trousers, and is armed with his 3G Technology mobile phone despite the lack of reception across much of the farm. Desks run along two walls of the small office reserved for him and Benjamin, the workshop storeman who helps him with paperwork. The room is dominated by a collection of grey, chest-high filing cabinets containing records of present and recent workers. On the opposite side of the room is a door leading to a hallway, which in turn leads onto the rest of the workshop offices where whites work: Willem, his wife, his brother-in-law, and the secretary (the daughter of a neighbouring farmer). While they all come through this door to seek out Michael or Benjamin, the latter rarely enter the rest of the office block. Michael and Benjamin work in close proximity to farmers and administrators. But they are overwhelmingly confined to responding to rather than initiating interactions. Nevertheless, such proximity ensures their easy access to the farmers in the eyes of other workers.

Michael's primary responsibility for much of the year is renewing permits. He maintains the paperwork and occasionally travels to Beitbridge border post to negotiate with both the Zimbabwean and South African sides. This requires not only knowledge of the various documents, but also the relationships and expertise to negotiate the bureaucracies at Beitbridge. This is of vital importance in a heavily policed border area, where Zimbabweans without papers risk deportation. Further enhancing Michael's reputation, the job involves a high degree of literacy. This both resonates with a widespread valorisation of formal school education – itself the result of Zimbabwe's colonial and post-colonial class history – and harks back to some workers' happier days in higher-status work than agricultural labour.

Though conspicuously solitary compared to Marula, Michael also sits at the interface between white and black, especially during the harvest. Enlisting the help of other workers, Michael moves the filing cabinets to the packshed and takes on a greater range of responsibilities. With a small team of other workers, he supervises harvest-time recruitment. He takes the new recruits to Beitbridge to arrange their visas. Throughout the harvest he handles paperwork for fruit buyers, and prints destination labels for the fruit pallets, work he refers to as 'data input'. The computer and buyer paperwork are in Willem's own office at the packshed. So what other workers see is Michael, dressed in an official-looking white coat and wielding a clipboard, spending hours each day in a room with their distant employer.

Michael shares Marula's proximity to the white bosses, but his closeness is of a different kind. Michael is not a lynchpin in the hierarchy in the sense of a daily avenue of communication between bosses and workers. His position confines him to narrower spaces of work-related responsibility. Consequently, he relies on formal channels to emphasise his seniority. One is his ability to hire and fire through explicit, private appeal to Willem. Another lies in the paperwork itself. Faced with large numbers of workers who need permits, he can exercise a degree of choice in whose he processes, and when. Especially during the harvest, when there are hundreds of recruits and many never receive legal documents, handling paperwork confers a great deal of power. But this power is limited when Michael is confronted by workers who have influence of their own. After Michael withheld the paperwork for one female worker, her husband assaulted him in his office. The latter was confident of André's support to prevent his dismissal. When Michael is challenged like this, turning to the white farmers themselves can bring further complication. In fact, Michael has been assaulted several times in the past, but he rarely reports such incidents because making a public issue of them would diminish his own sense of authority.

Despite such occasional assaults, however, compound residents continue to see Michael as a figure of high status and connection to the bosses. He is understood to be a man of great influence. His monthly pay, once equal to Marula's, subsequently became the highest among black employees. Since Michael's work is highly unusual in the black workforce, it is known that Willem sees him as indispensable and takes his complaints seriously. But Marula is more central to the day-to-day labour process in which most workers are involved. Meanwhile Michael's authority comes from his responsibility for records to which others lack access, through interactions with a powerful boss which, since they occur behind the doors of offices, others are unable to observe.

Models of Hierarchy in the Compound

At one level, this is a rivalry between chief supervisor and office worker, one that appears time and again in studies of work (see, for example,

'Management' or 'Paternalism'? 169

Beynon 1973; Burawoy 1979). But in a South African resident farm workforce, the contrast has particular consequences that extend far beyond the working day. In particular it can be seen in contrasting residential arrangements. Not only do workers live in close proximity, at their place of work; housing is also basic and much of compound life takes place outdoors. As in the classic case of the mines (see Gordon 1977; Moodie with Ndatshe 1994), relationships forged in the workplace spill over into compound life more generally (for Zimbabwean farms see Rutherford 2001). At Grootplaas, senior workers buttress their work-related status in the compound, and their behaviour in the compound mirrors that which characterises their work. Residential arrangements reflect notions of responsibility and authority.

Marula's dwelling reflects this. Centrally located, it is a focal point of the compound along with the farmer-owned shop. As the longest-serving and most securely established black worker, he has a brace of two-roomed senior workers' houses, and therefore his own toilet/shower block. He has built his own compound around this double-house, with a yard marked by low mud walls, a wood-covered porch, a mud/wood outhouse, a driveway, a fire-heated boiler, and underground piping to his shower for hot water. In front of the house is a treehouse for his many children. But, most importantly, underneath the tree is a seating area for drinking and holding court, with a second row of lower benches for busier occasions (see Figure 6.2). Here, the *khoro*[3] sits to judge minor disputes among compound residents, such as adultery cases. Led by Marula, other elders are largely men from his Lands team who become picking supervisors during the harvest.

It is not only in the highly institutionalised *khoro* that Marula wields influence. He is widely known to help those who ask. For example, he takes in particularly conspicuous new arrivals. Two female teachers became seasonal workers. He felt they would suffer in noisy seasonal accommodation, and had them share a room with his daughters. Looking after such highly educated women buttressed his status as protector. Similarly, a healer stayed with him for several months, and Marula meanwhile availed himself of his services to secure protection from invisible attack. The lines are blurred that distinguish an act of personal, individual assistance from one involving resort to the *khoro* or to Marula's worktime, foreman's role. For Marula is often to be found on weekends relaxing in his seating area with his drinking partners, many of whom are the court elders. At other times he cruises around the compound in his *bakkie*: occasionally recruiting men for unexpected jobs; more often just surveying the lie of the land. But whether as part of an institution built around him and his home – the *khoro* – or in his own right as foreman,

[3] Court of compound elders.

Figure 6.2. The yard of Marula's house, with the *khoro* on the right.

Marula dispenses judgement and assistance. The generalised authority of his daytime work extends around the clock and into the compound, and he embodies a distinct model of ethical responsibility to subordinates. Marula is seen by residents not only as representative of white bosses, but also as a figure capable of exerting considerable leverage. As a lifelong resident of the farms, and with numerous kin on estates across the area – more on this below – he stands for the relative stability and embeddedness that mediates the transience and violence of the border.

Marula's idioms of authority, and his and others' expectations of his position, are mirrored in less pronounced ways by other senior men at Grootplaas. As discussed in the previous chapter, the men who belong to Marula's Lands worker team have occupations as picking supervisors that are inseparable from their positions as local notables. Pickers are keen to establish good relations with them to ensure beneficial treatment outside as well as during work. Many supervisors sit on the *khoro*. Meanwhile one, Hardship, in his mid-twenties and still too young for the *khoro*, is nevertheless closely associated with Marula. He is responsible for the recruitment process and the daily register for pickers, both of which offer opportunities for preferential treatment. He also organises a seasonal-worker football team, which furnishes further occasion to dispense favours. The team members, being temporary workers, share accommodation and lack secure places for their monthly wages. Hardship and his coach look after their earnings, storing them in their own, lockable, permanent-worker rooms. Hardship goes further, even allowing them to use his vegetable garden. His high status in the compound is boosted through his role as organiser of high-profile gigs at the farm, status further reflected in his leather wide-brimmed hat and jacket. For Hardship, as for other supervisors, his conspicuous influence during and outside work establishes him as both figure of authority and provider of assistance.

The status of Marula and his supervisors is buttressed by residents' appeals for assistance and judgement. Michael sees his responsibilities rather differently. He is known in the compound as telling others to sort out their own problems: 'You've got your job, I've got mine.' Matters are of course more complicated. Michael does form personal relations with and helps particular people. When impressed by a newcomer's education, he may act to ensure an offer of employment in the packshed, which amounts to indoor work with higher pay. But, reflecting a more sharply bounded notion of occupational authority, Michael's assistance to others in the compound is confined to dispensing highly individualised favours.

While Marula's well-visited house suggests an extension of family life into the compound, the appearance of Michael's senior-worker house underlines his privacy. The outside is strikingly unadorned: a bare yard (see Figure 6.3) is marked by a high hedge and contains only a low table for washing up, a disused vegetable garden, and a crumbling mud-and-pole shed built by a previous occupant. Even in comparison to most other senior workers, Michael's yard looks basic, even neglected. This is not the case inside the house. Unlike other residents, he has bought a double bed and sprung mattress, and owns not only a large television but also hi-fi separates and floor-standing speakers.[4] With these he entertains guests, either sitting inside or watching the television through the door. Sometimes, he moves one speaker into the yard, pumping out loud music in the manner of the compound's beer sellers, declaring his resources to all around. But only a select few – the men Michael is drinking with at the time – are welcome to join. At other times, he brings guests into the house and shuts the door, an unusual practice in the compound. Michael's privacy limits discussion of his non-work past. Keen to highlight his professional success, he keeps details about the rest of his life to himself.

---

[4] Unlike Marula's investments, most of Michael's are potentially mobile. During my fieldwork, Michael was in the process of buying building materials for a home, which he would subsequently complete, next to his brother's (mother's sister's son's) house in Beitbridge District in Zimbabwe. In his late forties, he was keen to establish a home beyond the farm. Marula, a South African born onto a farm down the border road, has nowhere else to go except his mother's house in Musina, 50 km away. Most residents, like Michael, have alternative accommodation beyond the farm.

Figure 6.3. Michael's yard.

For example, though he has several adult and school-going children, my fieldwork was well advanced before I even heard mention of them.

While Marula's lifestyle is intended to place him at the centre of residents' existence, Michael's marks his managerial seclusion and sophistication – his difference from mere farm workers. Idioms of management and paternalism are played out even in workers' living spaces. But the managerial model is valued only by Michael and a few other residents. And the mode of sociality it promotes – with its sharp distinctions between manager and rank-and-file, and between work and leisure-time responsibilities – leaves him isolated.

Michael's situation is all the more difficult because he has incompatible goals. For, despite his aspirations to managerial sophistication and distance, he is nevertheless keen to assert his importance in farm dwellers' non-work lives. Drawing a sharp distinction between work and leisure time, he turns to alternative pastoral roles, consistent with his self-understanding, which he has accrued through the interventions of a British supermarket's development foundation. This is an organisation intended to address the historical disadvantage of South African farm labourers, in accordance with government policy to promote Black Economic Empowerment. Michael's education and current administrative occupation make him an obvious gatekeeper of the supermarket fund. His task has been to represent and make evident that which farm workers have the potential to become – 'empowered' and given skills.[5]

Michael was appointed the senior facilitator at the foundation-funded Adult Literacy Centre, known by compound residents as 'the school'. Perched at the top of a small slope, above the compound shop and the senior workers' houses, the school mainly taught computer-based courses in English reading and writing. Certificates, accredited by a well-known South African educational institution, were issued on completion of a proficiency level. Offering educational advice, overseeing work sessions, marking homework, and chasing up course participants were all tasks that enabled Michael to extend an educated style of authority into the compound. This was a source of status that he was more than willing to mobilise in his disputes with Marula, as he bitingly suggested to the latter that he come to the school to learn some English. However, attendance wavered, no one completed a course, and eventually the programme was discontinued. Workers, who feel that they are stuck working on a remote farm, often have difficulty seeing the point in such qualifications. Michael's teaching responsibility, though a way to occupy a pastoral role while underlining his difference from other workers, appeared similarly peripheral.

The supermarket foundation has offered Michael a further opportunity to accrue authority outside work time. During the period of fieldwork, he gradually became established as a spokesperson for the farm, representing it in meetings with the foundation. He was subsequently chosen by the foundation as a 'beneficiary' member of its Board of Directors. This meant attending meetings across South Africa and assessing applications for project funds from other foundation-affiliated farms. His position has signalled his administrative responsibility and status well beyond the farm, and his difference from his co-workers. He has held meetings in the compound to gather requests for projects that he takes to the foundation, underlining a quasi-paternal role as a dispenser of foundation wealth. However, as with his teaching position, these efforts have met with limited success. Workers feel that, as in his work for the school, Michael is too keen to underline his difference and distance from other workers. This detracts from his fulfilment of the generalised responsibilities that come with seniority. At one meeting, Michael was shouted down, and the meeting abandoned, because it was felt that he was making decisions about workers' welfare that were not his to make. He had begun organising the establishment of a games room and of land for workers' gardens.

---

[5] The names of the supermarket and foundation are left anonymous here.

Despite the obvious utility of both, it was objected that he had not consulted his fellow compound dwellers. It is in part because of his failure to find a central position in compound life consonant with his managerial aspiration that Michael asserts his status through stark seclusion.

We have seen how the occupations of the top black workers at Grootplaas contrast, and how they each relate to wider work processes. Their roles spill over into the wider life of the compound, with each presenting himself through different notions of work-based responsibility and status. But to grasp what is at stake beyond simply different jobs requires us to ask what Michael and Marula represent at Grootplaas. What each man evokes either helps to constitute or challenges interpretations of contemporary white commercial farming. For these occupations are expressed through contrasting models of seniority. To understand the conflicting registers, we first turn to the perspectives of the white farmers.

Paternalism versus Managerialism: Farmers' Perspectives

Formally retired from the daily running of the farm, Koos has a clear awareness of what he has achieved. He emphasises not only how he cleared bush and planted the area with other men of his generation, but also talks of his previous experiences – agricultural, commercial, and military, from the Free State to Kenya. As he spoke to me during my teatime visits to his house, it became clear what being a farmer meant to him. Crucial to this was a fundamental difference between 'European' and 'African', the former characterised by far-sighted vision, the latter by a level of instinct that Europeans have lost through exposure to 'civilisation'. His self-understanding as a farmer was that of a visionary, pioneer, and embodiment of modernity, a figure of guidance (see Chapter 2). Koos sees Afrikaners in particular as experts on 'the people', in this context meaning black labour.

As noted in Chapter 2, Koos employs a romanticised stereotype of 'the African'. Other farmers in the area would speak in a similar manner with similar enthusiasm. His traditionalist representation extends to the Grootplaas work hierarchy. Koos gives a particular gloss to Marula's role in the workforce, to the way he holds court and looks out for the interests of residents in the compound. Despite his involvement in smuggling and game poaching, Koos said, 'we support Marula as an *induna*. That's why he drives the newest *bakkie*.' The term *induna* suggests a headman-like idiom of authority, evoking earlier labour structures on mines in the region. But claiming that 'we *support* Marula as an *induna*' goes further, suggesting that his authority is merely sanctioned by the farmers. This is not the description of a relationship with an employee so much as an

'Management' or 'Paternalism'? 175

invocation of indirect rule. Certainly it is a long way from the managerial idiom on which Michael relies.

Koos is unimpressed by Michael, and scornful of his display of education. As Koos tells it, things were tense between them from the outset. When Michael was employed, he insisted on speaking English, not Fanakalo or an African language, and a violent confrontation was narrowly avoided. Koos revealed the importance of this point, when he claimed that his own rapport with workers owes much to his ability to 'speak the lingo'. Michael's insistence on speaking English disrupted Koos's self-presentation as a particular kind of farmer who understands African workers, speaks their languages, knows their stories, and wants to keep them in their place.

Koos sounds a lot like the classic image of the white farmer, with his self-understanding as pioneering visionary, and his image of Marula as a chief-like figure. Koos's son-in-law, Willem, illustrates a contrasting depiction of farming. Willem is concerned to present the Grootplaas enterprise as a profit-driven organisation like any other, rather than as a project of white vision and paternalism. Michael's position – as a black employee who is the corporate-style personnel manager, not a manual worker – enables Willem to emphasise this version of agriculture. It was no accident that Michael was the first worker to whom I was introduced, by Willem, at Grootplaas. Willem also prefers to see a sharp separation between work and home life. Feeling that he has only a limited grasp of dynamics within the workforce and the 'cultural' bases of disputes among workers, he speaks of his reluctance to intervene in non-work matters.

Unlike with Koos, the proper place for me to spend time with Willem was in his office. During one such meeting, Willem set out the Grootplaas labour hierarchy, using very different language from Koos. He first described the company's four shareholders – Willem himself, Koos, and two of Koos's sons. He then moved on to senior employees: there is the (white) workshop foreman and mechanic, and André, the production manager, who handles everything to do with growing the fruit. Also part of management are Michael, personnel manager; Marula, general manager; and the (black) packshed manager. These three sit on the management side of the table in discussions with workers, and are salaried.

The way this organisational mapping contrasts with Koos's understanding appears to herald a transformation in the nature of farming. Such a contrast may indeed derive partly from generational difference, but it also reflects different backgrounds. Willem, now in his late forties, was born in Rhodesia, the son of a school headmaster. Before he married into Koos's family, he had trained to be an electrical draughtsman. Crucially, moreover, differences in perspective between Koos and Willem are a reflection of their different roles now. Koos used conversations with me to express a romanticised account of the enterprises and successes characterising his life. For this, he drew on an equally romantic African backdrop. For Willem, speaking to me, a researcher, was more of an exercise in public relations. As the man responsible for representing the farm to buyers, supermarkets, and government, he was keen to highlight how – far from the oft romanticised or vilified popular image – farming is about running a production-oriented business. Speaking of the 'management side of the table', and the three black employees sitting there, Willem painted a picture in which management and labour are key polarities, as in any industrial operation, but race is not.

Within this new understanding, Michael is a key figure, since he is the one black employee who is not a manual labourer. Willem is keen to promote him as evidence of a businesslike outlook at Grootplaas. In conversations with me, he was careful to underline Michael's organisational importance, something he did not do in the case of Marula. Michael's directorship at the British supermarket development foundation lent Willem's assertions greater plausibility, giving a new reality to the figure of black manager, and underlining a willingness to adapt at Grootplaas.

Michael's occupation appears to fit better with the younger Willem's account of farming, Marula's with that of the older man, Koos. But a non-racial, managerial version of farming barely extends beyond the office. Despite Willem's managerial rhetoric, Grootplaas is run according to highly racialised conceptions of proper pay, accommodation, and modes of interaction. Revealingly, when Willem described the farm's labour hierarchy, he listed white before black managers. With Michael, Willem adopts a commanding tone he would not use with whites: words slowly articulated and a hard, monotonous edge to his voice. Whereas with André, the white production manager, he discusses progress in the orchards over an afternoon cup of tea in the tea room, with Michael his interaction is kept strictly within the office. In this regard, Michael remains unambiguously a black worker, separated by the racial divide reflected in the farmhouse/compound distinction. This leaves him with a kind of status that, as we shall see, proved extremely fragile in the years after my fieldwork.

Willem himself draws on both idioms – management and paternalism – depending on the situation. In conversation with me he emphasised the former. But he speaks to workers about his role as protector of his 'people': the police, for example, must speak to him before interfering with the farm's population. While complaining that the minimum wage and accommodation standards make it difficult for them to sustain off-the-books generosity, he and other local farmers talk about their continuing pastoral obligations towards workers: transport to hospital; flexibility about work leave according to workers' needs; rewarding loyal

'Management' or 'Paternalism'? 177

service through employment into older age and providing jobs for relatives. Some offer free transport to football matches. For the HIV/AIDS peer education team's Christmas party, Willem's son, Jacques, shot a warthog for their *braai*.[6] The farmer's wife may send small birthday gifts to the children of senior workers. The farming family also donates cast-offs to the workforce: the presence in the compound of items like their unwanted dartboard are evidence not only of the farm's sharp inequalities, but also of farmers' self-conception as benevolent patriarchs.

Taken together, the perspectives of Willem and Koos suggest managerialism at Grootplaas to be little more than rhetoric, a way of casting the status quo in a new idiom while establishing Michael as a symbolic black manager. Neither is managerialism as a description of actual work dynamics widely shared among most established workers. I heard comments criticising Willem's sharply bounded conception of work as being unusually abrupt. Meanwhile, Koos's own self-understanding receives support. People comment on his racism, but also praise him for his fluent SeSotho, in which he chats and jokes. Koos's version of Marula's position – one involving a vaguely defined, generalised authority, legitimised by workers, cast in a headman-like idiom – is broadly accepted. Koos's negative opinion of Michael as an ill-fitting member of the workforce is similarly reflected among workers. Although Michael's role as supermarket foundation director has been important for the farm's external image, on the farm itself it has cast him as different from everyone else, leaving him isolated.

This difference, however, is one Michael himself promotes. Willem's need to present Grootplaas as on a trajectory towards corporatisation resonates with Michael's own aspirations. Michael and Marula are not merely shaped by the views of their employers. Each promotes his own version of seniority and status. Exploring their self-understandings reveals how managerial and paternalist ideals of labour hierarchy at Grootplaas refract concerns related to Zimbabwean migration into South Africa.

Managerialism and Paternalism: Conflicting Registers of Status

Michael came to the farm in 1997, when Zimbabwe's economy was beginning to suffer but before the exponential increase in emigration. Like many more recently arrived Zimbabweans, he lacks a lifelong farm background and is keen to avoid having people draw the conclusion that he has sunk to the level of farm labourer, an occupation so demonised in Zimbabwe that many farms there historically employed Malawians or

---

[6] Barbecue.

Mozambicans (Rutherford 2001). He characterises his past in terms of an education-centred narrative, marked by the accumulation of formal qualifications: O-levels, and diploma courses in Personnel Management, Office Management and Administration, Computers, and Modern Management. His parents hailed from a town in southern Zimbabwe, but he was sent to stay with his mother's sister in a rural area. The reason was to ensure that he avoided urban distractions and concentrated on his schoolwork at a church-run day school. For secondary school, he returned to town. When asked in an interview why he left school, he cited financial constraints, like many other respondents. But he also added that he wanted to spend his time doing courses relating to his office work, that of till operator in a bakery. At 18 he entered the army for 15 years, finishing as an administrator/supervisor. After a brief stint driving a taxi he had bought, he visited his brother (since deceased) at Grootplaas and found employment. As he recounts his arrival, he had to hide his true, highly skilled nature from workers to avoid hostile attention. Only when he was told there was no permanent labouring job available did he reveal his qualifications and secure himself a 'management' job.

Michael now speaks proudly of his position as 'manager of this organisation', employing distinctly corporate language. But he is also aware that his job can be framed as little more than that of a farm labourer, a feeling augmented by the contrast between his own fate and that of another of his brothers. In 2007, Saul,[7] a registered tour guide living in Midrand outside Johannesburg, visited the farm for the first time, arriving in his new double-cab *bakkie*. As we sat with beers in front of Michael's house, Saul told Michael how he, his elder brother, had been a role model growing up. But now he was disappointed: for Michael had become a mere farm worker. Michael validated his position by saying that people here need leaders, but the conversation had touched a nerve.

For those, like Michael, who are disappointed with their fates but have attained a senior post on the border farms, a managerial idiom can serve to maintain a sense of self-respect. Benjamin, the storeman/clerk with whom Michael works, feels a similar ambiguity. Michael emphasises his authority as manager, rather than mere farm worker, by establishing difference from everybody else: through his administrative work, connection to the British supermarket, proximity to the white bosses in their office, and private life in the compound, replete with high-status consumer goods. This is one way of dealing with a problem faced by an increasing number of Zimbabweans: how to assert their histories of education and class aspiration, now lost due to economic collapse at home and consequent geographical dislocation. No doubt Michael feels all the

---

[7] Actually his mother's sister's son, but his mother's sister has taken on a motherly role since he lived with her, and especially since his own mother's death in 1996.

more acutely the need to be not 'just' a farm labourer as he seeks out and engages the best educated of the new arrivals.

As the previous chapter illustrated, the class dimensions of this distinction are inflected by ethnic ones, mobilised together in the construction of stereotypes. The core workforce at Grootplaas is Venda-speaking. But those seasonal workers who have higher-status backgrounds tend to be Shona, or Ndebele like Michael. Marginalised within Zimbabwe, Venda tend to have had less access to educational opportunities. Their long-standing association with South African farms is interpreted as acquiescence within anachronistic and racialised agricultural hierarchies. Seasonal workers with higher-status backgrounds assert their difference in ways that mirror Michael's. They refer to management books to criticise work dynamics and the dominant notions of authority and responsibility in the workforce. They characterise themselves through education and etiquette. And they tell stories of home that emphasise access to commodities like cars. The few couples granted their own accommodation invite guests inside their rooms rather than socialise in public.

Michael's style is impressive to some Grootplaas workers with middle-class aspirations. But they are seasonal recruits who move on. In any case, Michael remains too much of a permanent fixture at the farm for high-status seasonal workers to identify with him. Meanwhile, a reaction to this same influx of relatively well-educated Zimbabweans buttresses Marula's position within the more influential permanent workforce. He has accrued widespread respect and status among permanent workers, especially TshiVenda speakers like him, after long years working on border farms. Marula was born on a nearby estate in 1959. His father and two brothers were foremen, his three other deceased siblings farm workers. Although some siblings attended school, Marula never had the opportunity to do so because his parents lacked the money at the point when he might have been eligible. He advanced slowly, beginning as a shepherd, then progressing to gardener, mechanic, and driver; eventually, he became a foreman. At Grootplaas, he can still point to old boundaries, and the sites of former compounds such as what is now the football pitch, where his daughter was born three decades ago. Speaking of his achievements, Marula emphasises his experience. Sons, daughters, other relatives, and old friends from the farms visit regularly. He is well-placed to cast himself in the classic patriarchal role.

Increasingly Marula is faced with recruits who draw on alternative bases of status to which he lacks access: education, urban sophistication, even non-Venda-ness. He is quick to point out that, while he may have no schooling (like the handful of other South Africans in the workforce), he *does* know farming. But this makes it all the more important for him to underline his centrality and superiority within the world of the farm. In this he receives support from senior workers who hail from rural areas just across the national boundary fence. They, like Marula, have attained powerful positions at Grootplaas, but are made to feel like yokels by some of their subordinates. From their perspectives, a previously Venda-dominated area has become overrun by non-Venda. Farm hierarchy is given new meaning as it confronts assertions of superiority based on personal histories far from the border farms.

Conclusion

Literature on South African farm labour explores the characteristics of paternalism, as a benchmark against which to measure its waxing or waning, and hence the transformation of labour relations. The resulting impression is of undifferentiated workforces experiencing change from above. This resonates powerfully with a popular regional view of farm workers: rag-clad masses on remote, semi-feudal estates. This chapter and the previous one have shown how farm workers are, in contrast, highly differentiated in class, status, and ethnic terms.

The value of situational analysis like Epstein's (1958) lay in exploring how particular relationships and events refract wider processes. At Grootplaas these are on different scales: a personal rivalry between two workers refracts farming's racial micropolitics; the border and Zimbabwean displacement; the South African political context; or supermarket corporate social responsibility agendas. These intersect, creating both allegiances and rivalries. In a period of uncertainty in agriculture, different visions of farming – an established paternalist model and a corporate managerial one – are taken seriously for different reasons. Both Michael's and Marula's styles of status and seniority are more than what they seem, and can only be understood through workers' reactions to these broader visions and the processes they entail.

Focusing on two top workers at Grootplaas, we see two visions of agriculture not only coexisting, but actively constituting each other as opposites through their rivalry. Both Marula and Michael appear invaluable. Pressure to corporatise comes to farms through particular farmers, like Willem. But they moderate their outlook selectively in the light of anti-corporatising paternalist ideals. And a 'modern', even 'neo-liberal', hands-off approach leads them to avoid actively engineering change *within* the workforce. Indeed, long after my departure, Michael's narrow dependence on Willem's good graces came into sharp focus (see Chapter 8). The combined effect means that farmers present one face to the outside and another to the workforce. Meanwhile, workers produce or challenge hierarchy, and its ethical entailments, themselves. Transformation involves all these, and is thus far from merely planned or top-down.

But Grootplaas lies at the intersection not merely of layered processes, but also of contrasting personal histories. While agrarian transformation itself is not simply cumulative and linear, we must also attend to how actors construct their own cumulative experiences. This requires transcending classic situational approaches. Epstein's focus on actors' current, situational roles risks being accused of presentism, and begs awareness of how people see the present in terms of their pasts. Understanding Grootplaas residents' roles involves relating them to their criteria of success in life. Marula's position as foreman represents the culmination of his gradual rise through farm-worker ranks. Michael's managerial position is the proper endpoint for his accumulation of qualifications. Their dispute is intractable precisely because each man's success frames the other as a failure. Koos sees *his* life as that of the classic white pioneer. His and Marula's interpretations of their pasts reinforce one another. At the same time, Koos's imagination of farming – which entails white visionaries and traditionalist Africans – undermines Michael's educational achievement. Michael's self-understanding challenges Koos's sense of accumulated expertise about 'the people'. The meanings of paternalism and management are shaped, not just by Grootplaas's various relationships, but also by the collision of its residents' historical experiences.

This chapter and the previous two employed a close analysis of dynamics in the Grootplaas workforce. They revealed how established, structured arrangements among residents at the farm both reflect and are produced by wider processes – changes in South African commercial agriculture and the effects of the Zimbabwean crisis. The next chapter broadens the scope of the book while still taking a situational focus on Grootplaas, by examining how the economy of the border and of remittances is structured by farm life.

7 Scaling Up

The Farms and the Border Economy

Introduction

As noted earlier, Southern Africa's highly regulated hubs of migrant labour have seen fragmentation and decline. Many people have turned to informal livelihood strategies. Consequently, according to Jens Andersson, the 'combining of migration and trade indicates that these movements can no longer be defined narrowly as *labour* migration' (2006: 376, emphasis in original). Andersson (with a focus on Malawian migrants) argues for a shift 'from a focus on economic centres and production relations towards the sphere of economic circulation' (ibid.: 394) – in other words, from work to trade. Taking seriously the point that migrants move for different reasons according to different patterns, however, means exploring the intersections between different forms of human mobility in the region. These include an intersection between movement oriented around trade and that arising from continuing labour migration. We need to attend to how traders and their businesses shape, and are shaped by, the experience and economic arrangements of waged working populations.

The small-scale trade at issue is what is often referred to as 'informal'. Informal economic activity is most simply defined by its extra-legality: 'income generating activities that take place outside the regulatory framework of the state' (Castells and Portes, in Meagher 2010: 14). Following scholars' recognition that state-sanctioned activities and illegal ones often interpenetrate, and may in some cases not be distinct at all, the usefulness of the term 'informality' has been called into question (in Roitman 2004, for example). What doing away with the term leaves unexplored, however, is how informality is constituted in opposition to formal work in particular settings. How, in other words, are small businesses, lacking state regulation, built around centres of waged employment? How are they different from the waged sector, and how do they articulate with it?

This chapter employs the analysis of the Grootplaas resident population offered in earlier chapters to make sense of the wider economy of the Zimbabwean-South African border. Earlier chapters demonstrated how *mapermanent* achieve provisional stability in compound life, and how more marginal farm dwellers relate to and largely depend on them. The book now turns to the relationships between border farms' settled populations and small-scale business along and across the border, demonstrating their interrelations. At Grootplaas, formal and informal livelihoods constantly permeate and constitute one another. The border farms represent islands of relative security on the border. White farmers' control of their land is relatively predictable, in part because it can be negotiated through senior black workers, and it mediates the sometimes unpredictable regulation of the state. Further, living in the border farm communities can offer a degree of everyday stability, unlike life in crisis-ridden Zimbabwe to the north, and in contrast to the uncertainty of South Africa's cities. For those employed on an ad hoc basis or staying on the farms through personal connections, business ventures bring in the regular income that enables them to continue residing there.

There has been a particular lack of attention to the relationships between formal employment and informal trade in the literature on border economies. In anthropological research, borders are often seen as sites of opportunity in the face of marginality, where state attempts to define spatial boundaries are perpetually confounded by creative enterprises and brokerage. Formal places of work are largely absent from such accounts. Border zones are often sites of activities – smuggling, local 'vigilante' justice, unauthorised movement – that are officially illegal but have become accepted features of everyday life for resident populations (see, for example, van Schendel 2005). From the perspective of border dwellers, there is an important distinction between 'formal political authority' – a legalistic, top-down view – and 'non-formal social authority' – local mores that are illegible to the state (Abraham and van Schendel 2005: 19). This distinction is key to understanding the murky realities of border areas. Making sense of economic activity in the eastern Democratic Republic of Congo, which relies on connections across international borders, MacGaffey (1991) highlights the role of the 'second economy', a similar concept to 'informality'. Creative entrepreneurship seeks out the interstices of state regulation. Keen to show how border areas are more fluid, and the state more contested, than often assumed, anthropologists of borders highlight local arrangements and activities that pass beneath the official radar. In doing so, they explore how illegibility from the view of the state offers opportunities to those in the know. Flynn notes that:

Differences in national economic policies, regional resources, and monetary currencies make borders lucrative zones of exchange and trade, often illicit and clandestine. Smuggling occurs across borders around the world, providing an important means of livelihood for border residents and prompting creative social networking and cross-border ties within borderland populations (1997: 313).

In this view, then, borders are shadowy places, a far cry from the structured arrangements of on-site workforces. There is some truth to these characterisations, borne out by my own observations of the Zimbabwean-South African border. The military are trained in bush tracking by a local game farmer to hunt down Zimbabweans who have come through the fence. They are concerned to catch not only 'border jumpers' seeking work in South Africa, but also smugglers. Smuggled goods include precious stones from Zimbabwean mines and contraband cigarettes in bulk. Gangs known as *magumaguma* operate along the border, robbing and raping those attempting to cross. State officials themselves make the most of their location. Reportedly, underpaid Zimbabwean police and soldiers also extract wealth from passing migrants. South African soldiers are known by Zimbabwean women to accept sexual favours in return for letting them go if they are apprehended.

However, on the banks of the Limpopo, the routine dynamics of workplaces and the businesses of informal traders are connected in important ways. A narrow focus on the more furtive side of borderlands risks skewing and misrepresenting them. While the existing focus on clandestine activities casts border dwellers as perpetually preoccupied with resisting state regulation, Zimbabwean-South African border dwellers make the most of their location by being visible to state officials in a specific way – as waged farm workers. This legitimates their presence, leaving them free to pursue a range of business ventures. Understanding this, in turn, reveals how the border farm workforces shape, and are shaped by, local informal business.

It is not only on borders that informal economic activity, associated with entrepreneurial creativity, is starkly contrasted with the drudgery of formal employment. As in the sharp distinction that is drawn between that which occurs 'above' and that 'below the radar' in the anthropology of borders, informal economic activity is approached as qualitatively different from waged employment because of the resourceful entrepreneurship involved. What the case of the Zimbabwean-South African border reveals, however, is the tight intertwining of waged work and small-scale enterprise. Arrivals at Grootplaas seek waged agricultural employment as a foothold in building more lucrative businesses; they take their money-making strategies with them as they enter waged labour forces; and they use their incomes to trade goods back home to Zimbabwe.

As the first part of this chapter shows, in many cases such businesses are not so much cases of moonlighting; they are, especially for non-permanent workers, the ultimate goal of farm employment. Their ventures allow them to keep their lives on the farm afloat – a relatively secure island of employment and dense social relationships. Indeed, although some farm businesspeople make considerable profits, others receive far more modest returns – enabling them to merely get by. What they share is a desire to stay on the farm. The second part of the chapter takes a wider view of the border farms, demonstrating further how waged employment establishes workers as traders. Although this book takes a workplace as its focus, the latter part of the chapter brings Zimbabwean workers' remittances into the analysis. For when they remit, many workers' practices lie at the interface between wage labour and informal trade. Remittances often take the form of goods for resale, as they seek out further opportunities to make a living in an uncertain and volatile economic environment at home. Striving for a degree of stability and predictability under unstable conditions, displaced Zimbabweans are in this sense both labour migrants and traders.

By unpacking the material and temporal articulations between waged work and other means of making ends meet, this chapter reveals border residents' wider projects to achieve a degree of everyday security through their workplaces. At the same time, it shows the farm compounds to be hubs of the border's economy, illustrating how resident workplaces act as magnets for diverse livelihood activities. First, however, I discuss the meanings of economic 'formality' and 'informality', and their particular regional inflection in Southern Africa.

Space and the Informal Economy in Southern Africa

The official compound shop is located in the middle of the Grootplaas compound. It opens onto the compound hall, a raised concrete area without walls, shaded by a corrugated metal roof, where meetings are held. At weekends, men play pool or table football in the games room; dice or draughts on boards chalked onto the concrete floor of the hall. More generally, when the shop is open it is a place for passing the time of day and catching up on gossip with Esther or Lindiwe, the two black women who are employed permanently as shopkeepers and live in the compound. Even during the working week, there is often someone – an off-duty worker, an unemployed person, or some compound children – leaning on the shop's fold-down hatch. During the period of fieldwork, the shop was managed by Koos's daughter-in-law, his son Paul's wife.[1] But she rarely entered the compound and, like other white residents at Grootplaas, when she did so she drove in and spent little time outside her four-wheel drive. A diverse stock is kept, catering to the compound's daily needs given residents' limited access to town. Frozen chicken, canned foods for relish, sauces and spreads, milk, painkillers, washing powder, soap, toothpaste, sweets, and soft drinks (cans or bottles): this gives some idea of the range. Stock is sold at prices slightly higher than those charged

---

[1] He was one of the white farmers at the time, but later ran a guesthouse some distance away (see Chapter 2).

in Musina. Airtime (credit) for pay-as-you-go mobile phones is sold at the official price and is an especially popular purchase. In South African terms, the shop is much like a lot of general stores.

During the harvest, maize meal is distributed from the shop at bulk-purchase rates on credit to seasonal employees who choose to participate in the scheme. In this respect, seasonal employees are dependent on the farmers for their staple food through a farm-bound paternalist arrangement. The money is docked from their wages. But this is no classic company store, supplying workers with the lion's share of their daily needs while getting them into debt. Except for this maize scheme, the shop requires immediate cash payments. This and its rigid, timetabled opening hours discourage potential customers, who turn towards the more flexible world of compound business. Residents sell groceries, beer, and soft drinks from their rooms, on credit and at all hours. There is a slightly higher mark-up on goods than at the compound shop, but profit margins are not large. Men who own *bakkies* drive along the border road to the official border post at Beitbridge and to Musina. There is a further plethora of enterprises, from pot making to haircutting, baby sitting to clothes mending. These businesses, by offering credit, allow residents to handle the constraints of monthly wage rhythms.

Meanwhile, on the first Friday of every month – payday at the border farms – large numbers of women travel to the farming area to sell goods to newly paid workers. With waged work paid at approximately the South African minimum wage,[2] those who receive such wages working on the farms represent a lucrative opportunity. Taken together, and in the context of widespread unemployment in South Africa, the farms make up an unusually large, and unusually constant, wage-earning population. Seasonal workers, unable to risk or afford going to town, are especially good customers. Most market traders are Venda and hail from different places in or near the nearby former Venda homeland, or from Musina. Farm sales are part of the monthly regional circuit, with stops timed to coincide with differing paydays. They stop off at schools when teachers are paid, at offices and police stations after pay, and at town markets.

From the perspective of farm residents, these traders satisfy demands for spikes of consumption immediately after pay. For many workers, payday is the time not only to pay off debts to businesspeople in the compound, but also to splash out on new items of clothing, food, and non-perishable groceries like soap. Some purchases are for use in the compound, others for remittance.

Understanding small-scale trade of this kind in relation to labour dynamics requires discussion of the meaning of 'the informal economy'.

---

[2] R989/month in 2007. Workers are paid either by the hour, which provides them with this monthly minimum, or, in the case of pickers, according to a piece-rate calculation that leads to underpayment.

For Keith Hart, who set the terms of more recent discussions, the formal/informal distinction, at one level a matter of whether or not transactions were subject to state regulation, was also a distinction 'between wage-earning and self-employment. The key variable is the degree of rationalisation of work – that is to say, whether or not labour is recruited on a permanent basis for fixed rewards' (1973: 68). Wilson formulates a version of this distinction as follows:

> The surplus labour force must create employment for itself in order to survive. This self-generated employment, marked by a qualitatively different mode of production from that employed in the formal sector, constitutes the informal sector. (2005: 38)

This relationship between different forms of livelihood has been interpreted in different ways. The informal economy has been variously valorised or condemned for its difference from organised production: seen as pre-capitalist; as a reserve of potential wage labour and the epitome of capitalist fragmentation; or as peopled by budding entrepreneurs (Fernandez-Kelly 2006). While earlier research saw the informal sector as something that would dwindle with expanding capitalist production, more recent analysts have shown this to rest on false, teleological assumptions. Informalisation, far from disappearing, is a worldwide trend with Africa on its frontline: 'the informalising dynamics of contemporary Africa represent an extreme in a wider global trajectory of rapid informalisation' (Meagher 2010: 14). In Zimbabwe and South Africa, informalisation is the result of different processes, both economic collapse in the former (see Jones 2010) and globalised neo-liberalism (see Bernstein 2007).

As Fernandez-Kelly (2006) notes, what is required is attention to how informality itself is shaped in particular settings. One approach is to consider the priorities and capacities of different state institutions, since formal work comes within state purview. With regard to borders, this has meant close attention to policing, smuggling, and brokerage. The on-the-ground realities of state regulation can lead to the view that the concept of informality is useless and misleading. Taking this position, Roitman (2004) discards the concept of informality outright. She shows how the cross-border trade of guns, diamonds, contraband petrol, and other extra-legal goods serves to strengthen rather than challenge the local state in the Chad Basin. The trade she observes is therefore not 'beyond the state' or 'antistate' (ibid.: 192).

This rejection of informality is a sign of the times. As Meagher remarks:

> Amid processes of deregulation, globalisation and weakening states, informal forms of economic organization have become so pervasive, and so deeply intertwined with formal economic structures that the old notion of an 'informal sector' or 'informal economy' has been called into question. (2010: 11)

However, the complete collapse of the formal/informal distinction leads to a loss of analytical purchase on real differences among economic activities. Arguing that differences have become blurred or non-existent leaves analysts little room to examine the actual processes involved in informalisation (ibid.).

Where informality is acknowledged, one danger is essentialising its character. In much research, as illustrated by the border literature discussed above, informality is seen to be characterised by creativity, adaptive agency, and entrepreneurship. Such creativity may even spill over, outside narrowly economic strategy. In *Congo-Paris*, MacGaffey and Bazenguissa-Ganga (2000) describe traders from Congo-Kinshasa and Congo-Brazzaville to Paris, who contest legal, spatial, institutional, and moral boundaries through their activities. They 'demonstrate an oppositional counterhegemonic culture in the ostentatious, competitive consumption of their lifestyle' (ibid.: 7). Informal traders excitingly shape worlds that are invisible, or only partially visible, to state officials. For some analysts, indeed, this shadowy world is an especial feature of Africa, where 'smugglers, diamond diggers, currency traders, fraudsters and simple migrants all find ways of evading laws, frontiers and official exchanges rates' (Bayart 2000: 260).

What is required is analysis of informal practices with sensitivity to the particular 'legacies', 'linkages', and 'localities' that shape them (Grabher and Stark, in Meagher 2010: 24–5). What this means is attention to the institutional context, with its changing history. There is more to this than figuring out how the state sees, or what reinforces state power. If the formal sector is taken to mean officially regulated waged employment – as it is by Hart and others – then what demands investigation is the nexus of control by *both state officials and employers*. Preoccupation with entrepreneurial exceptionalism draws attention away from close analysis of the ways unregulated business and law-bound waged employment constitute one another. The usefulness of the notion of informality lies in leaving room to explore this articulation. In border research especially, preoccupation with traffic through the fence may lead informality to be conceptualised narrowly, primarily in relation to illegal trade. This focus examines less attentively how such practices relate to, and are interwoven with, organised production. The case of the Zimbabwean-South African border enables us to move away from accounts limited to cross-border traders, to explore how 'wage-earning and self-employment' (Hart 1973: 68) articulate. How do businesspeople and waged workers differ in such a setting?

Rather than occupying sharply delimited spheres, 'formal' and 'informal' economies are intimately related to one another. Indeed, their relationship can take different forms. Informal activity can be conceptualised as what happens *beyond* the rules of formality. This is the perspective taken in much of the literature discussed above, and is the version that leads Roitman (2004) to contest the validity of the distinction. But informal activity can also be read as *complementary* to formal arrangements, or even as *necessary* to their function (Hart, personal communication).³ The case of Grootplaas invites such a reading. The fact of white farmer control of the land leaves the formal black workforce stranded in the middle of the countryside in compounds, with limited access to goods and services. These are provided informally, by people who rely – for their residence in proximity to customers – on connections to the farm through employment and influential friends. Informal services are what enable workers' lives at the farm, and therefore what underpin the farm's formal production.

Using the formal/informal distinction to make sense of people's livelihood possibilities and their connections has a spatial inflection in Southern Africa, and particularly in racially segregated centres of wage labour. The region's history has produced particular notions of informality, formality, and the relationship between them. Since 'informal' refers to unregulated activity, its meaning has been associated with racialised apartheid- and colonial-era policies that controlled black movement and residence, often in the service of capitalist interests. Resident labour forces have a long history. Historically, black labourers on the mines lived in compounds – fenced and tightly controlled (Gordon 1977; Moodie 1994). More generally, black movement, settlement, and residence was controlled under South African apartheid and in pre-independence Rhodesia, in part by designating townships characterised by regularity, austerity, and residents' lack of any permanent rights of tenure (see, for example, Ginsberg 1996; Barnes 1999; Lee 2005). In many Southern African settings, formal employment was historically defined as working for whites, and had a spatial dimension because of corresponding access to accommodation.

It is therefore unsurprising that, as Preston-Whyte (1991) and Bozzoli (1991; Bozzoli with Nkotsoe 1991) show, many women who moved to cities during the apartheid era pursued livelihood strategies that were defined in opposition to employment by whites: that is, to their own domestic work or their husbands' waged jobs. In the former case, residence in white homes offered a base from which women could establish businesses such as clothes repairs and liquor brewing. Such women, as in compounds on the mines (see Moodie with Ndatshe 1994), established a range of businesses to satisfy black demands for (sometimes illegal) goods and services, in the interstices of white spatial control.

Studies of informal black businesses in white-designated spaces have focused on the mines and on domestic work. Meanwhile, Rogerson contends, 'in South Africa, home-based enterprises, including the running of

---

³ Thanks to Keith Hart for pointing out this distinction.

backyard workshops, hairdressing salons, shebeens, or *spaza*[4] operations, have long been a hidden feature of the urban scene' (1991: 336). White farming areas have not received the same attention in discussions of the South African 'informal sector'. And yet farm spatialities share features with the spatial orders of domestic work and those of urban township businesses. On the one hand, black residents live on private, white-owned land and therefore, like many domestic workers, depend on their employers for residence. On the other hand, however, the compound is a zone of relative autonomy, where the farmers keep their distance and where black residents openly establish various businesses. Although police and army personnel regularly patrol the farm compounds, they are willing to enter into agreements with farmers about the status of their 'people', even if the latter are undocumented. Being tied into the spatial logic of the farms and their compounds ensures that residing on the border and establishing businesses there are possible. At Grootplaas, the most important conceptual distinction between ways of making a living is not legal (grocery sales, for example) as opposed to illegal (such as smuggling or alcohol sales). Rather, it is farm employment as opposed to other ways of making money while resident on farmers' land.

'Illegality' is marked out in some business settings by its risks. But on the Zimbabwean-South African border, the risks of illegal trade are relatively low. Residents do not foreground distinctions between 'legal' and 'illegal' goods, except in so far as they affect the risk of *kusungwa* (being arrested). While sellers speak of having their beer confiscated by police, I never saw this happen, and it is sold and consumed publicly in the relatively secluded farm compound. Cigarettes are easily shifted and sold in small quantities. And although inspectors occasionally come to check for smuggled meat – the problem here is foot-and-mouth disease – a far greater risk is being caught poaching game from the estate by the farmer. Despite concerns about large-scale smuggling of cigarettes and precious stones into South Africa, farm populations bring goods through the fence for consumption in the residential labour compounds without regulation. In the process, precious stones are sometimes also conveyed. Border-patrol preoccupation with 'illegal' migrants is what isolates seasonal workers, whose fear of arrest turns them into lucrative, captive consumer markets on the border farms. But for businesspeople themselves, crucial challenges stem from the relationship between waged work and their own enterprises.

Could all of this not be interpreted simply as a case of workers moonlighting to supplement their wages? Hart (1973), in his study of economic opportunities in Accra, notes the importance of moonlighting to complement daily earnings, and shows people juggling different sources of

---

[4] Small, informal, retail shop, often run from seller's place of residence.

income from both official and unofficial economies. As one would expect, there are many such cases at Grootplaas. Gregory, for example, is a senior permanent employee who works the water pumps in the orchards. Outside his working hours he sets up his Singer sewing machine under a tree in his yard, and plies a brisk trade repairing clothes. His goal, he says, is to live on his income from tailoring so that he can remit his farm wages to keep his children in school.

However, the distinction between farm employment and other ways of making money while resident on farmers' land is characterised by complex mutual dependences, and requires further unpacking. This is more than a case of moonlighting. As we shall see, farm work *itself* cannot be fully understood without appreciating that for many it keeps open valuable informal economic opportunities, some of which bring in more money than agricultural employment itself. Meanwhile, the form remittances take in the context of economic crisis in Zimbabwe blurs the lines between formal employment and informal trade. Formal employment may therefore be seen as a key resource in wider strategies of business establishment, material accumulation, and attempts to establish predictable lives on the farms.

Making Money at Grootplaas: Spatial and Temporal Articulations

'I am a businessman,' announced Chipo triumphantly. Dressed in a suit and wielding a wad of thousands of South African Rands, he was preparing to leave Grootplaas for his home in Zimbabwe following the end of the 2007 harvest. He had worked the picking season in the packshed, loading crates of citrus onto pallets. But the money in his hand was not his pay packet from the farm. He had spent the previous evening rushing around, tracking down the large number of residents who owed him for the loose Madison cigarettes he had been selling throughout the picking season. Bringing cartons across the border from Zimbabwe, he had begun selling cigarettes at Mopanekop, the neighbouring farm, in 2000. He returned there to continue selling during the harvest most years until 2006, when he came to Grootplaas and was employed seasonally, first as a picker and then in the packshed. This was his second year working at Grootplaas, and his cigarette business had flourished. Having stable accommodation during the harvest, a place where customers could find him, as well as a permit (he was one of the few seasonal men to receive one) all helped his enterprise. His case is illuminating. For his cross-border sales preceded waged employment, which then further enabled his money making. It did so by allowing him a degree of stability, and by securing him a work permit. This meant that police and soldiers, concerned with 'border jumpers', would leave him alone. Waged employment left him nevertheless feeling fundamentally a businessman, as his sartorial transformation and his explicit declaration, just mentioned, attest. It was a resource in consolidating his enterprise. It made him visible to state officials in a particular way: as a legitimately employed worker.

In some cases, the link between farm work and wider money-making strategies is not merely a matter of secure residence. For some men, it is precisely their employment that offers business opportunities. Consider Daniel, a tractor driver who dealt *mbanje* (marijuana) while working. The *mbanje* is brought through the fence by women, in plastic cups that are a standard unit of sale. His job involved hauling full trailers of oranges from the picking teams to designated places in the orchards, from where other drivers would take them to the packshed. And he would bring empty trailers from here back to the picking teams. There was a lot of waiting, within close range of pickers, offering ample opportunities for clients to approach him. This was essential to his business. For *mbanje* is consumed by men as a work stimulant. He could not have sold as much had he been a foreman, say, in the steady-paced, conveyor-belt world of the predominantly female packshed. His close proximity to competitive, aggressive male picking teams was the ideal setting for his business.

Selling cigarettes and marijuana are at least mobile activities. For these, farm employment is useful in embedding sellers in the working population. But many people attempt to establish more enduring enterprises, such as a *spaza* (grocery shop) or shebeen (bar). Chipo would have been unable to found anything of this sort because of the seasonal character of his employment: his residence lasted only five months. Daniel had sold beer in the past, but had stopped, possibly because illness left him increasingly weak and tired.

Longer-term enterprises develop through stable residence, and at the same time contribute to their owners' rootedness in the Grootplaas population. Consider the case of MaiJimmy (the mother of Jimmy), Daniel's sister. She sells beer from her room, drawing weekend revellers with her sound system. Soldiers, as valued guest customers, sometimes sit inside her room on stools to drink and talk while she shuttles in and out serving beer to sweating Saturday night dancers. But she also sells salt, sugar, light bulbs, toilet rolls, soft drinks, yoghurt, and other products from Musina.After buying a fridge and a single crate of beer for resale in 2005, she soon expanded to *braaing* (barbecuing) small fish called *bakawaya*[5] during the harvest and then to a wider range of stock. Like other *spaza* and shebeen 'owners', MaiJimmy relies on her room in the compound. She lives and trades there all year round, but to do so she must work the picking season in the packshed, and remain on call for odd jobs as a 'semi-permanent' employee throughout the year – one of around 20

---

[5] Described to me in English as mackerel.

such women. Of course, farm employment brings in some of the money with which she supports several siblings' children in Zimbabwe – now including those of her brother, Daniel, who died in 2007. But, whereas she used to sell 'part-time', alongside more extensive formal employment, she has shifted her primary emphasis to her sales, as they bring in more cash. To understand the importance of MaiJimmy's farm work, one must grasp that what she mainly needs is continued access to her room, in which she can keep and sell her stock. And as one of the established, 'semi-permanent' women at the farm, she has a work permit, which offers her immensely greater security. Meanwhile, the income from her *spaza* enables her to stay at the farm throughout the year, despite the fact that she gets only an irregular income from the farm itself. Both forms of work have allowed her to build a relatively stable, predictable life at Grootplaas among people she knows well. The short-term exchange of her business transactions has contributed to her long-term place in the social life of the compound.

It is less easy for men to maintain a similar balance between wage work and trade. This is because the 'semi-permanent' employment category is exclusively female. Such women represent a reserve of labour to clean the compound and guard the orchards against baboons. They are also retained to mitigate the almost entirely male composition of the permanently resident workforce. Men are either 'permanent', in which case they work full-time throughout the year, or 'seasonal', in which case their employment and their status as residents ceases in September or shortly afterwards. Permanent workers do sometimes run businesses like shebeens – Daniel did so before his illness – but such enterprises are necessarily undertaken on a part-time basis. Where trader men do benefit from the residential security afforded by work on the farm, it is not their own but their wives' or girlfriends' semi-permanent status that enables this. Full-time male taxi drivers and *spaza*/shebeen owners use their female companions' accommodation as a base. Josiah, for example, is a former employee, but because of his girlfriend's continued employment is permitted to stay in the compound and help her run their store. From her room, they sell both beer and groceries. They also attract dancers and gamblers who can listen to booming music under the light rigged on the wall outside. This gathering point is an ideal place to cook and sell *makwinya* (fried dough balls) – as they do with the assistance of Guidance (Josiah's brother and a picker). Thus its grocery store role is overshadowed by its emergence as a major social spot, a shebeen that rivals MaiJimmy's establishment on the next row. The room is a crucial base for income-generating activities, not only for Josiah and his girlfriend, but for his brother Guidance too.

Businesses like MaiJimmy's are supplied by informal taxi services operated by men who own *bakkies* and wait for customers at a clearing on

Figure 7.1. Park Station, the compound car park.

the compound's edge, next to the border road (see Figures 7.1 and 7.2). Basically a lay-by marked off with old tires embedded in the ground, the compound's car park is known jokingly by some as Park Station, after Johannesburg's central transport hub. Given the farm's distance from town, it is through these informal taxis that Grootplaas residents mitigate their dependence on their employers. Permanent workers ride them to town for shopping trips after payday, and traders depend on them to bring in new stock. Exploring one taxi driver's experience further illuminates how the nexus of compound enterprises operates.

Cornelius was born and grew up on a farm down the road, the son of Zimbabweans although his father became a naturalised South African. He himself holds dual citizenship. He worked for years as a driver for a delivery company based at Polokwane, three hours to the south, using his brother's car. On one trip to Grootplaas he met his wife, who works at the packshed. He started from that date to take regular drives up to the farm when he had time off work. While doing this, he would offer lifts to people travelling between the farm, the border post, and Musina. He quickly realised that he could make much more money by doing this than in his delivery job, and moved in with his wife at Grootplaas. He switched to driving the border route, still in his brother's car, in 2005. He already had some savings, but began to build on them by selling soft drinks from

Figure 7.2. Informal *bakkie* taxi carrying departing seasonal workers and their supplies.

his wife's room in the compound. A reflection of how lucrative such trade can be, he bought a small *bakkie* within the year. Cornelius now has a home at Makushu in the former Venda homeland, beyond Musina to the southeast. He goes there at month's end and holidays, but otherwise stays at Grootplaas with his wife, or in the compound at Swartvlei, another border farm, where a brother works.

Cornelius started small, accumulating start-up capital by selling drinks, a strategy typical of those adopted by businessmen on the border farms. Like them, he was able to secure the crucially important residential base in the compound. Most permanent employees are men but whilst, as noted, their jobs leave little time for well-developed enterprises, 'semi-permanently' employed female residents have both the permanent accommodation and the time to run well-established businesses. Men who wish to make a living from business in the compound must also have kinship or other connections with existing residents, which in turn are built through their business activities. Cornelius's experience reflects this. Although he now has a house in Venda, his enterprise continues to depend on residence in the border compounds, and is therefore enabled by the jobs of his wife and brother. His business is defined by its simultaneous dependence on and independence from the wage logic of the farms.

The conditions governing residence on the farm are of greater consequence for such informal businesspeople than their legal status. Regulation more often means sanction from the farmer than from the state. However, given the hands-off character of white farmers' labour control, this regulation is usually mediated by senior permanent workers. Farmers themselves have limited knowledge of or interest in the goings-on of the compound. Traders and drivers are vulnerable to the power of *mapermanent*, who have the ear of their employer – especially the most senior men such as Marula, Michael, or Hardship. In Cornelius's case, some workers resent his staying on the farm and making money from the workforce. During my fieldwork, Cornelius was told to leave by one of the farmers, having been accused of stealing petrol. It emerged, he said, that he had been set up and reported on by influential male workers who had attempted to make advances towards his wife. After being rejected by her, they attempted to get rid of the couple by having her dismissed from employment, thus denying both of them a residential base. Although Cornelius and his wife successfully convinced one of the farmers of the situation and were not ejected, the example shows how precarious their position is. It is not only that Cornelius's presence is dependent on his wife's employment and residence. It is also that, as a woman in the compound with no influential male employee to speak for her, his wife's own position is itself precarious, because of the considerable power of particular male permanent workers.

Like the 'informal' activities organised around domestic work in South Africa's cities, business possibilities rely on continued residence. But unlike in such situations, the continued right of residence at Grootplaas depends on mediated paternalism. Ultimately, the goodwill of powerful permanent workers is what spells the success or failure of informal businesspeople in the compound. They must ensure that they maintain good relations with the permanent workforce, and therefore they extend credit on very generous terms to *mapermanent* and, on more restrictive repayment cycles, to others. MaiJimmy, for instance, like other similar vendors, sells almost all of her stock on credit, writing customers into her account book and chasing them up after the monthly payday. MaiJimmy claims that she assesses customers according to need rather than trust. She sizes up their domestic obligations – how many mouths they have to feed – and makes concessions accordingly.

These credit arrangements supply traders with customers throughout the month, despite the fluctuations in available cash around the monthly pay cycle. But extending credit is also a risky practice. There is very high labour turnover among seasonal workers, and many leave without settling their debts. Traders attempt to mitigate this by waiting at the farm offices at payday, each with a brown exercise book in which accounts are recorded. Their aim is to catch workers as soon as they receive their money for the month. However, even this strategy is difficult to implement. It is hard to guarantee repayment even from well-known customers. Sellers do not seem to collaborate to blacklist customers who default on their debts. And businesspeople are low-key about demanding their money, too concerned to maintain good relations. The official compound shop, established merely to keep workers supplied with necessities, neither depends for its existence on a healthy profit, nor must it contend with the risks of credit or a precarious foothold in the compound. But traders and drivers have to negotiate their informality, in the sense of running non-farm enterprises on farm land, without enforceable sanctions. For them, flexible hours and generous credit policies are ways to ensure reasonably stable customer bases. Informal business relies on relationships with other farm residents. In turn, businesspeople become key figures in the compound.

Traders and taxi drivers – people like MaiJimmy and Cornelius – are mutually dependent in their precariousness. Drivers not only transport people; they also bring commodities to compound traders. Some charge a small fee. Others buy goods and sell them on to traders at a marked-up price. These arrangements are flexible. Drivers might go to Musina even without customers, in order to bring stock, losing money on petrol by doing so. At other times they may bring small packets of goods without charging a transport fee. Regular customers may have their luggage fee waived when they travel. These concessions are important because compound residents who make a living from selling town goods or providing transport depend on a degree of predictability. They need steady demand and a reliable customer base.

Diverse businesspeople ply their trades at Grootplaas. They do so in the face of residential and financial precariousness. But this ensures that they offer generous terms of credit and are prepared to work all hours. Their flexibility makes them crucial to the Grootplaas workforce. For the fact of monthly pay and consequent cash fluctuations in the compound would make life very difficult without credit. This flexibility both responds to the wage rhythms of the farm, including the rigid hours of the compound shop, and also defines the informal compound economy in opposition to them. The temporal articulations between waged work and informal trade are as crucial as the spatial connections. The constraints placed on farm workers' lives by their white employers offer opportunities to informal businesspeople. But the farms themselves can only function because black residents satisfy these workers' needs by establishing informal arrangements.

The relationship between waged work and informal business does not stop here. Businesspeople in the compound buy their supplies from wholesalers in Musina, and both their trade and that of the taxi drivers stem from the farm's isolation. For undocumented seasonal workers, especially, leaving the farm to go to town is a great risk. But in theory, any resident could begin selling goods he or she had bought in Musina, and many permanent workers do so on occasion, capitalising on the expense and difficulty of transport to town in return for a small profit.

In a parallel manner, Grootplaas workers – permanent and seasonal – make the most of the fact that they are in South Africa and can obtain basic commodities that are scarce in Zimbabwe. Migrants' remittances as members of a labour force require understanding in relation to their roles in trade networks. And seasonal workers especially – as unsure of Zimbabwe's future as of their own – attempt to leave their options as open as possible.

Sending Soap: Remittances as Business

During the period of fieldwork – midway through a period of acute hyperinflation and commodity shortages in Zimbabwe – employment at Grootplaas represented access not only to South African rands but also to a range of goods.[6] Sending such goods to Zimbabwe became a key means for workers to send their earnings home in a form that would hold more stable value than the hyperinflatory Zimbabwean dollar. Doing so also enabled Grootplaas residents to remain flexible in their investments. For their purchases could be consumed on the farm or sent home as needed, and used or sold there. Consumption in the compound, remittances from farm labour, and cross-border trade often blurred into one another. As Rutherford (2008) observes, such survivalist strategies occasionally become means for modest accumulation. Whereas the previous section showed how border trade is structured by workforce dynamics, this one shows how workers' economic strategies responded to the worst period of the Zimbabwean crisis. Here the focus of the present study on labour relations at a workplace is broadened to analyse workers' consumption in terms of their remittance practices. Doing so illuminates a final intersection between the established dynamics of labour migration in the region and those emerging through displacement and the navigation of crisis.

There are differences between *mapermanent* and seasonal workers in their use of earnings, reflecting the contrasting place of farm employment in their lives more generally, as explored in chapters 4 and 5. Underlying these differences, however, are similar efforts to hedge bets in the face of uncertain futures. Secure employment, and wages according to the South African hourly minimum, enable *mapermanent* to remit larger amounts of cash. Some remit steadily; others accumulate and then

---

[6] What I now describe was a direct result of hyperinflation and supply shortages following government-imposed price fixing in Zimbabwe – both phenomena specific to the period of fieldwork.

take leave. After one mid-harvest payday in 2007, a small survey revealed that 20 out of 49 randomly selected permanent workers had travelled home. Three sent their remittances with someone else, such as a worker from the same area of Zimbabwe, or through the services of a trusted bus driver at Beitbridge. Residents report that cash remittances regularly top R500 monthly, around half of pay, and may exceed R1,000 in the case of those with successful business operations. Beyond this, *mapermanent* juggle ways of spending their money. Many save for their retirement, acquiring housing materials and occasionally cattle. These larger investments are sometimes achieved through rotating credit groups, which emerge and disappear depending on their members' reliability. At the same time, *mapermanent* invest in life in the compound, as discussed in Chapter 4. Spending on housing, girlfriends, and beer, all of which root *mapermanent* in the compound, divert considerable resources from other projects. A third way of spending earnings became particularly important during the period of fieldwork, as hyperinflation rocketed and retail-store shelves in Zimbabwe emptied. *Mapermanent* would put a large proportion of their earnings – up to R500 monthly – into non-perishable groceries for kin at home.

This somewhat schematic account gives some sense of *mapermanent*'s competing priorities. What it does not do, however, is convey the extent to which *mapermanent* attempted to remain flexible in uncertain circumstances – to navigate the shifting terrain of crisis amidst a 'loss of coherence and unity' (Vigh 2008: 10) beyond the border farms. As described below, their use of commodities blurred the boundaries between these apparently different logics of expenditure, allowing particular purchases to be used for different goals as needed.

Among the seasonal workforce, pay according to a piece rate often leaves pickers with too little money to send home. Those hoping to find further employment in cities to the south must factor in the costs of transport. Moreover, whereas *mapermanent* are embedded in networks of other border residents to organise sending resources to Zimbabwe, many seasonal workers, as newcomers, lack the necessary connections. A brief stint on the border farms may leave little opportunity to arrange remittances. Nevertheless, many seasonal workers do send, or hope to send, something home monthly during their contracts. In my 2007 survey, 33 of 53 seasonal workers went home, and two more remitted through a wife or neighbour. But the amounts are much smaller than in the case of *mapermanent*. Instead, for many, end-of-harvest spikes in spending form the basis of their remittances. Mirroring *mapermanent* responses to the Zimbabwean crisis, however, seasonal workers' remittances at the end of their 2007 contracts combined cash and commodities. Like *mapermanent*, they attempted to keep their options open, albeit with fewer resources than their core-workforce counterparts.

Figure 7.3. Monthly payday market in Grootplaas compound.

A starting point for understanding workers' attempts to maintain flexibility is the compound's payday market (see Figure 7.3). At the end of each month, the compounds of the border farms are abruptly transformed by large numbers of saleswomen. The market at the end of the final payday of the harvest is especially large. At Grootplaas, it covers a whole section of the compound around the crèche and adult literacy centre. Women arrive packed onto informal *bakkie* taxis, lugging 'Filipino suitcases': enormous, tartan-patterned, woven nylon bags – large enough for an adult to fit inside – full of stock. By late afternoon this section of the compound is covered in plastic sheets, on which diverse goods are laid out. There are clothes, new and second-hand. Food on sale is both fresh – such as tomatoes, apples, and potatoes – and ready-cooked, including hard-boiled eggs and deep-fried chips. *Mashonja* (mopane worms)[7] are sold dried, both long-lasting and ready to cook. Traders also sell music cassettes. Especially popular are non-perishable groceries, such as Omo clothes-washing powder and its brand rival, MAQ. Workers are usually paid during the afternoon. As they filter back from the workshop, the market area fills up with potential customers who wander among the displayed offerings – browsing, comparing prices, and snacking on

[7] Caterpillars of a large moth, dried for sale and fried or stewed as a relish.

Scaling Up

Figure 7.4. Seasonal workers leaving the compound after the harvest, at dawn.

the deep-fried chips, apples, or hard-boiled eggs being sold. At dusk, traders light candles or turn on battery-powered standing lamps. When trade finally wanes, well into the night, the traders sleep on their plastic sheets, next to their stock.

In 2007, departing seasonal workers at Grootplaas (see Figure 7.4) bought large quantities of goods from the market for remittance. These included clothes, but groceries predominated. Trips by taxi to Musina, en route to the border post, were further occasion to procure supplies (see Figure 7.5). There was little fear of deportation among those already on their way to Zimbabwe. In Musina, one man bought a DVD player, at his wife's behest, as well as a new case for a mobile phone he had bought second-hand from another departing worker. Chipo, whom we met earlier, bought a two-foot-tall amplifier. Both made their purchases at a shop where a former farm worker from the border area now worked. But far more popular – and reflecting spending in the compound – were the wholesalers where compound traders themselves buy their wares. The Musina trips were, in effect, a means to circumvent the middle-man, since they involved very similar purchases to those made at the market. Departing workers spent very large amounts of money on groceries. Chipo and his wife, for example, bought a vast quantity of different

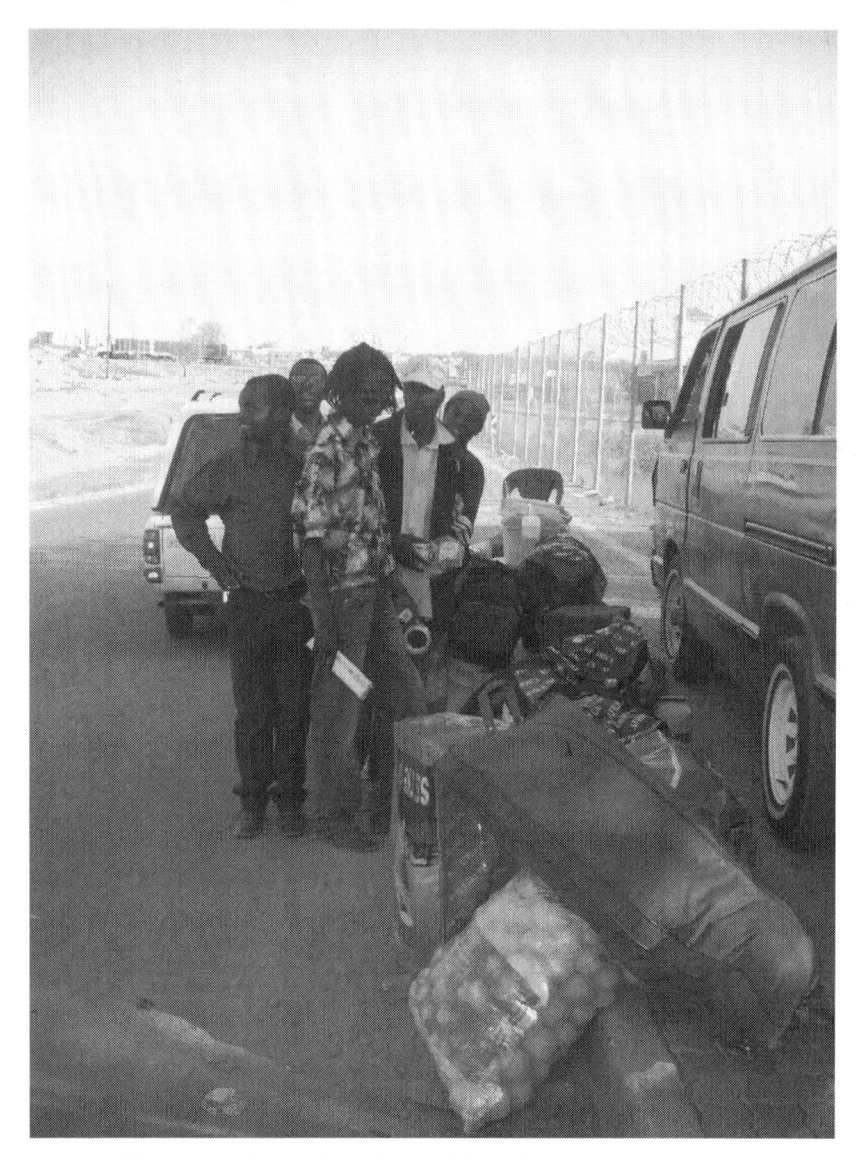

Figure 7.5. Seasonal workers waiting with luggage at the border, after a trip to Musina.

items – biscuits, corn snacks, a box of green-bar laundry soap, a bag of bath soaps, rice, flour, and a number of other things. The total cost was R665, the larger part of a monthly wage at Grootplaas.

These spikes of spending reflect wider choices about consumption, oriented towards remittance. Both permanent and seasonal workers send both cash and commodities home to kin. Their ad hoc remittance

Scaling Up 203

strategies contrast with the model suggested by classic accounts of regional labour migration from the 1970s and 1980s (for example, Mayer 1980; McAllister 1980; Murray 1981), which focused chiefly on the gold mines. In those cases, male labour migrants in the region remitted wages with the very particular goal of building up their homesteads. Like their contracts for employment, their remittances appeared highly structured. Social mechanisms strongly encouraged labour migrants to avoid spending their wages at their workplaces and to send home the cash they earned. Indeed, in some cases – Mozambican miners, for example – this was formalised through a system of compulsory remittance or deferred pay. But there has also been a long regional history in which migrants sent home a range of commodities in a more ad hoc manner, to address supply differentials between places of work and home. Zimbabwean women have sold crotchetware and cigarettes in South Africa and returned with in-demand goods at least since the 1980s (see Zinyama 2000; Muzvidziwa 2001). In the early 2000s, Mozambicans in Mpumulanga Province, South Africa, would take home whatever commodities were needed (Norman 2005). During acute hyperinflation and supply shortages in Zimbabwe, Grootplaas employees often sent their remittances in the form of diverse goods – a picture that defies a simple structural account. Remitting was often more a matter of keeping kin and *musha* (ChiShona: rural home) afloat through a present shortfall than building a longer-term future through investment and expansion.

This cannot be seen as an unprecedented response to the Zimbabwean crisis, given the well-established practice of using migrant earnings to send home needed items. But just as Zimbabweans responded to Economic Structural Adjustment Policies by expanding and adapting existing patterns of cross-border trade in the 1990s, so those who worked on South African farms between 2006 and 2008 reoriented well-worn practices of remittance-in-kind to address contemporary economic troubles at home: the range of goods they sent, including basic staples, reflected the widespread shortages there.

Zimbabwean farm workers engaged with hyperinflation through particular strategies of remitting goods. As economic anthropologists have shown, in any social setting there are categories of wealth that are not exchangeable for one another (see Bohannon 1955; Gudeman 2010; also Parry and Bloch 1989). Distinct categories of value may be especially useful for labour migrants facing the perennial problem of earning wealth, sending it home, but not being around to control its use. Ferguson (1985, 1992) explores the remittance strategies of male labour migrants from rural Lesotho. They invest in cows, purchased with cash, that can only be sold – re-exchanged for cash – under dire circumstances. By buying cows and defending the one-way barrier between livestock and money, these men attempt to hold value in a form protected from the everyday material demands of kin at home.

But Zimbabwean labour migrants on South African border farms have done the opposite, responding to economic crisis at home by avoiding investing in goods not easily re-exchanged there. Rather than sending goods whose use and exchange can easily be controlled, they have hedged their bets as a response to hyperinflation. As noted, almost all sent both goods and cash, in somewhat varying proportions, but it is the choice of goods that is most revealing. Permanent workers, especially, would invest in things like livestock and housing materials, building up their homesteads. But most non-cash remittances took the form of small, everyday commodities that could both be used at home or resold there as needed, for Zimbabwean dollars that were necessary but lost value at an alarming pace. In the context of hyperinflation-induced price fixing, and consequent supply shortages, non-perishable goods had particular value. Cooking oil, for example, was hard to acquire in Zimbabwe for a while, and consequently Grootplaas workers would send it home. The large quantities of body cream and washing powder sold at the compound market on the last day of the 2007 harvest were not just for use in the compound. By sending goods with different possible uses – either easy exchange or non-conversion – Zimbabweans on the border navigated their way through crisis and acute uncertainty back in Zimbabwe.

By bringing goods to Grootplaas and its neighbours, therefore, the monthly market traders were not only supplying commodities for daily use. They were also delivering supplies that could be remitted. Indeed, with certain items, shortages in Zimbabwe offered the possibility of taking home a non-perishable good in bulk, and selling it off as cash was needed. Green-bar soap – each piece around a foot long, and used to wash clothes – is particularly well suited to transport. The long oblongs do not decompose and come in neatly packed wholesale boxes of 24. One way to hold wealth when remitting, therefore, was to become a trader oneself, albeit on a very small scale. The market traders who come to Grootplaas therefore sell to people who extend trade networks into Zimbabwe in response to distribution problems there. The farm represents a nexus in a regional, informal-sector supply chain, because it is the source of waged employment.

The crucial point here, however, is that soap and similar commodities lent themselves to *different* uses. I heard both personal use and resale cited as reasons to remit such items rather than money. But remitters did not need to decide beforehand. This was important given the potential risks, for those without connections, of being caught selling rands for Zimbabwean dollars on the black market.[8] Similarly, goods bought for

---

[8] In September 2008, Zimbabwean Central Bank Governor Gideon Gono licensed certain shops to sell goods for foreign currency (*Mail and Guardian* 2008). After that, Zimbabwe officially became a multi-currency economy, within which US dollars and South African rands could be used. At the time of my fieldwork, however, rands could not be used legally in shops.

use in the compound itself could also be used for remittance. Just as soap and cooking oil offered flexibility back at home, so possessions at the farm defied sharp distinctions between goods for the workplace and those for remittance. Many at Grootplaas feel themselves to be in exile from economic circumstances in Zimbabwe, where their earnings rapidly declined in real value during hyperinflation. Such purchases – hotplates for cooking, blankets and crockery for comfort in compound rooms, warm fleece jackets for early mornings at work – come to operate as another way to hedge bets. People establish lives at Grootplaas with items that can also easily be taken home and used there. Attempts to establish the conditions for congenial domesticity, described in Chapter 4, do not preclude later re-use of some resources beyond the farm. This is all the more important, given the uncertain future of Grootplaas itself, to which I turn in the concluding chapter.

During the period of fieldwork, Zimbabweans at Grootplaas did not target particular forms of wealth that kept spheres of use and exchange sharply separated. Quite the opposite: they kept their options open with goods that blurred such distinctions, a response to the unsettling instability of hyperinflation. Their decisions spoke of a future with uncertain horizons. Better-off workers on the border farms did fit the classic pattern of Southern African labour migrants, incrementally building long-term rural futures through remittances. But even they devoted substantial resources to negotiating crisis, keeping life back home sustainable in the present.

Appreciating Grootplaas employees' flexible approach towards commodities – eliding work use and home use, personal consumption and trade – enables us to understand their responses to hyperinflation as part of a wider reorientation of the border's economy. In turn, it once again calls into question sharp distinctions between wage work and other ways of making ends meet on the border, and between labour migration and displacement amidst crisis.

Conclusion

Anthropological literature on borders often focuses narrowly on the clandestine informal economy. By neglecting wage work in such settings, authors promote a sharp distinction between informal and formal economic activity that ignores the ways workplaces (the world of 'production') relate to their surroundings (the world of 'reproduction'). By celebrating the below-the-radar creativity of informal entrepreneurs, without exploring the articulation between such businesspeople and formal wage work, organised production is often marginalised from analysis. Reflecting this, Andersson (2006) neglects 'productivism' – a focus on labour relations – in his account of migrant trade, with which I began. This chapter has examined informal economic activities in a fresh light, by exploring their articulation with wage work. What Grootplaas illustrates is how a workforce both structures and is structured by the local informal economy.

On the Zimbabwean-South African border, a range of goods and services is key to the economic life of the area's agricultural estates. White agriculture, to keep turning a profit, relies on diverse informal efforts to meet basic needs in workforces. In turn, some link with secure employment at the farms is imperative if one is to establish business activities. Jobs ensure accommodation, and recognition of employment by soldiers and police ensures a degree of security. Traders either hold agricultural jobs themselves, or live with others who do, calling into question any sharp distinction between the workforce and the various enterprises that surround it. Formal and informal livelihoods not only coexist, but also directly constitute one another. Meanwhile, understanding the logic of consumption at the farms' monthly markets blurs boundaries between workers and entrepreneurs, as farm employees often plan to resell goods in Zimbabwe as a response to economic uncertainty there. This has included working as small-scale traders in response to acute commodity shortages at home. From one perspective, the Grootplaas compound is full of businesspeople. But these various forms of work and exchange are all ways in which Grootplaas residents, surrounded by transience and uncertainty, reorient their practices to establish provisional stability. In the concluding chapter that follows I turn to the theme of uncertainty itself, exploring Grootplaas residents' prospects as the farm's future is called into question.

8 Conclusion
Between Production and Fragmentation

What might be the fate of the arrangements described here in the longer term? Two developments towards the end of my fieldwork had profound implications for the future of Grootplaas, and the fate of its residents. One was the Zimbabwean elections of March 2008. The other was the appearance of a research team looking for coal in the uncultivated parts of the border estates, as farmers considered selling their land to a mining concern. Each was the subject of intense discussion in the compound. Here were two sources of possible change in the lives of workers and their dependants. From the point of view of this study, both moments provoke wider reflection. Caught between fragmentation and organised production, what are the prospects of farm workers on the Zimbabwean-South African border?

Continuity and Change: The 2008 Zimbabwean Election and the Coal Prospectors

On Saturday 29 March 2008, the day of Zimbabwe's presidential and parliamentary elections, the excitement in the Grootplaas compound was palpable. Some *mapermanent* were wearing bright yellow T-shirts advertising Simba Makoni, an independent presidential candidate. The T-shirts had been distributed at the Beitbridge border post, and Grootplaas residents had acquired them when returning from a trip to Musina. Few farm dwellers, however, ended up crossing the border to vote. Many had been unable to return to Zimbabwe to register during the officially appointed period several months before. For others, the trip to vote was too expensive and meant taking time off work. There were also concerns in some quarters about political violence, and assumptions that the elections would be rigged as earlier ones had been – '*zvakangofanana*' ('it's all just the same') was one reaction I heard. Nevertheless, the results were eagerly anticipated by most. Discussions around the compound conveyed the hope – for some, the belief – that this election *would* be different to previous ones: that it would be fair and would bring a new government to power. One worker, in greeting me as he walked past, mimed a spanking action in the air. The symbol of the opposition MDC party is an open hand; the spank, I was told, symbolised 'the MDC effect' on ZANU (PF) in the elections.

Strikingly, given the political atrocities that had characterised ZANU (PF) rule, much talk of change was less concerned with the cessation of violence than with what one might expect in less volatile settings: the implementation of effective policy. 'Why is Mugabe saying what he *will* do; why has he not already done it or [why is he not] doing it now?' asked one man. Another expressed the matter by means of an agricultural metaphor: 'We need to change fields from a non-productive to a productive one.' Hope for a better Zimbabwe emphasised economic improvement, and many saw an MDC government as a way to bring back 'white investment'. Underlying all this, however, was an awareness of recent injustices, although this was still couched in relatively mild terms. Mugabe was not a 'tyrant', but 'a father living well while not supporting his family'. Support for Morgan Tsvangirai, the leader of the main MDC opposition party, was expressed in similarly low-key terms. He was a good prospect, not because of any sterling qualities, but because his recent arrest and assault proved that he *was* in opposition rather than being a collaborator in Mugabe's regime. Border farm residents' hopes for change thus focused on possibilities both for economic stabilisation and improvement, and for an end to the brutality of the current regime. But all their hopes were cautiously expressed.

Soon, however, the mood in the compound soured. The elections had been on a Saturday; by midweek there was still no announcement of the winner. Rumours about the imminent release of results, fuelled by media speculation, proved unfounded. Residents rushed in vain to their televisions and radios after work each evening. A week after the polls closed, MaiJimmy the shebeen owner, upset and suspicious, complained to her customers: ZANU (PF) was already talking about a re-run of the election, even though no official results had been announced.[1]

Workers' anticipation of change at home was replaced by their resignation about ZANU (PF)'s stranglehold on power. Circumstances appeared set to continue as before, and this continuity was underlined by the arrival of the annual harvest in the Grootplaas productive cycle. Once again, hundreds of prospective pickers climbed through the border fence, lacking other livelihood options. Among them, as before, were those who feared for their safety, but now with greater intensity. They had voted

---

[1] The two MDC factions were eventually recognised as constituting a parliamentary majority. In early May, a re-run for the presidency was declared, as Tsvangirai's lead over Mugabe was deemed too small to constitute an outright victory. In the June re-run of the elections, ZANU (PF) backed violent reprisals against opposition voters, and Tsvangirai consequently retired from the presidential race. A 'power-sharing' deal was struck, with Mugabe continuing as President and Tsvangirai as Prime Minister. ZANU (PF) retained control of state institutions such as key state security forces.

MDC and now faced a litany of violent acts by youth militia, state security personnel, and war veterans backed by ZANU (PF).[2] George, the picker described in Chapter 5, was one such fugitive. Grootplaas residents had not been unrealistic about the speed of Zimbabwe's hoped-for transformation; they expected the kinds of changes that would allow them to leave farm employment to take a long time. The damage done to the Zimbabwean economy, they often remarked, would take years to undo even with a new government, but at least a new government would mark the eventual end to Zimbabwe's turmoil. However, it was now clear that the political situation that underlay Zimbabwe's crisis remained in place. Residents now increasingly spoke of the power behind Mugabe of Joint Operations Command – the clique in control of state security forces – and the generals' own pecuniary interest in maintaining current conditions.

Their indefinite dependence on farm employment was what the elections – and the fact that they brought no change for ordinary Zimbabweans – underlined with stark clarity. Nor would the official adoption of a multiple currency system in 2009 – and the consequent end to hyperinflation – have a measurable effect on this dependence. As Worby (2010) notes, one characteristic of the Zimbabwean crisis from the point of view of displaced Zimbabweans is its open-endedness, its 'temporal uncertainty' and its 'indefinite indignity'. Competing with a sense of temporary exile, therefore, is one of 'abjection', the sense of 'debasement, hopelessness and socio-political ostracism' (ibid.: 431) that results from the collapse of previous notions of success in life and personal progression (see Ferguson 1999). For many of the more transient Grootplaas residents, in particular, there was no end in sight to a life of employment-seeking migration.

What made all of this worse was the gradual realisation that the future of the border farms themselves was anything but secure. Even *mapermanent*, it seemed, might soon join their seasonal counterparts as part of a floating regional reserve of labour. In the last six months of my fieldwork (from late 2007), rumours began spreading of the possible sale of Grootplaas to a coal-mining concern. It had long been known among farmers and *mapermanent* that there was coal under the farm, and some long-term compound residents had anticipated such a sale. But whereas in the past the farmers had merely prospected themselves, now an Australian multinational corporation was taking core samples from across the estates. As Chapter 3 showed, the border region was shaped in important ways by South Africa's twin pillars, mining and agriculture. Now, on the Limpopo River's southern bank, mining appeared to be edging farming out.

---

[2] See Chapter 1.

210   Zimbabwe's Migrants and South Africa's Border Farms

The uncertain future of the farm highlighted the contrast between the attitudes of *mapermanent*, those of the seasonal workers, and those of the farmers. *Mapermanent* had invested in the compound and had settled in its barrack housing. For them, Grootplaas was their home. Its decline would mean a loss not only of occupation, but also of residence and a web of social relationships: indeed, of the very 'hubs of employment' with their accompanying structured arrangements which this book has analysed. Seasonal workers were less rooted in life at the farm than *mapermanent*. But, for many of them, the farm was a point of stability in a continual search for livelihood. Not only a place of employment, it was also a hub of diverse relationships. It was a place to which they knew they could return each year for work, and where they could expect to see familiar faces. For the white farmers, as for their core employees, the farm was home. However, faced with a precarious environment for white commercial agriculture, farmers increasingly saw the estates in terms of business decisions. The prospect of land claims on the border farms had led them to view their land as an asset. If necessary, they would sell for as good a price as possible before moving on. Unlike members of their black workforce, the farmers could – and did – respond to insecurity by becoming more flexible capitalists with a portfolio of investments.

Faced with the possible end of life at Grootplaas, *mapermanent*'s responses were varied. One worker with a long history of employment, who had been living at the farm since before the present farmers bought it, assumed that residents would simply stay put and revert to subsistence agriculture, planting mealies (maize). After all, this was not only a commercial farm; it was also a home to black residents. A female compound dweller asserted a similar sense of the right to reside when she commented: 'If this mining comes here they are bound to look for other places for us because we have been staying here for a long time and we will not go anywhere. This is our home; they must build houses for us. Where will we go?'

Most *mapermanent*, as noted in earlier chapters, have homes in Zimbabwe where they can reside. But Zimbabwe's ongoing economic troubles, extending beyond the 'Zimbabwean crisis', mean that many such homes are places merely of shelter, not production or sustenance. As foreman, Marula is perhaps the most dependent on continued residence at the farm. While he has developed his accommodation into a fitting focal point of compound life, he has no home beyond Grootplaas except his mother's house in Musina's township. When confronted with the advent of the coal prospectors, his response evoked an experience typical of the classic South African farm worker – that he has the right to live and work at Grootplaas because of his long personal, familial, and working history there. Such claims, however, were unlikely to be observed by a

Conclusion 211

transnational mining corporation.[3] In all likelihood, if the border farms disappear, *mapermanent* will lose the various forms of status and stability they have secured in the farms' workforces, as they move from place to place in search of waged employment. For Marula, even more than for other long-term residents, the overturning of the fragile and delicate balance presently in existence on border farms would spell massive and unsettling change.

This story of two scenarios of change, with the threats and promises they contain, raises wider questions about the relationship between fragmentation and production in contemporary Zimbabwe and South Africa. Whereas this book has explored their intersection through a study of labour relations, it concludes by reflecting on implications for the future.

Fragmentation and Capitalist Production

The two aspects discussed above – Zimbabwe's recent crisis and the insecurity of South African agriculture – both have particular causes, though these are intertwined.[4] However, their *effects* have contributed to a wider process of fragmentation, as argued in Chapter 1. This kind of uprooting of populations produces a floating reserve of vulnerable work seekers, and has been seen as central to the creation of workforces under capitalism. According to this narrative, economic relations – including labour relations – are disembedded, disconnected from the various social ties of power and obligation in which people live. The employment relationship is depersonalised, and workers become atomised and alienated from one another. Does fragmentation in Zimbabwe and South Africa lead to a similar disembedding of economic relations in the interests of capital?

In *The Great Transformation* (1944), Karl Polanyi argues that fragmentation made possible the creation and intensification of capitalist production. Taking the paradigmatic case of nineteenth-century England as his focus,[5] he asserts that a market-driven capitalist economy

---

[3] South Africa's Extension of Security of Tenure Act 62 of 1997 (ESTA) should in theory ensure that established farm workers like Marula are protected against summary and arbitrary eviction, as 'ESTA occupiers'. Only evictions accompanied by a formal court order are legal. ESTA was designed precisely to address the fact that farms represent homes, not just workplaces, for many farm workers. However, ESTA's record has been extremely uneven, and illegal evictions of farm workers far outnumber legal evictions (see Hall 2003, 2007).
[4] See Alden and Anseeuw (2009) on the influence of Zimbabwean land reform on the South African process.
[5] Polanyi's narrative offers a way of thinking comparatively about the place of economic relations and wage work in social life. This requires a somewhat schematic view, which downplays the complexities of his historical account. A comparative and schematic use of Polanyi's insights is nevertheless invited by his own use of ethnographic and historical

depends – albeit unsustainably[6] – on markets in labour and land, which have to become commodities for purchase and sale. Disembedded economic relations mean breaking down forms of dependence – those of kinship, as well as vertical ties of paternalism – and replacing them with individualised contracts:

> To separate labour from other activities of life and to subject it to the laws of the market was to annihilate all organic forms of existence and to replace them by a different type of organisation, an atomistic and individualistic one. (Ibid.: 171)

Though not the result of deliberate policy to create a workforce,[7] the effect of Zimbabwean displacement has been that it causes Zimbabweans to travel – alone or in small groups – to places often far from home to work for wages. The experiences of seasonal workers on the border farms appear to mirror the atomising effects described by Polanyi. Migrants' insecure existence, bare accommodation, and merely provisional associations with one another turn them into conveniently malleable units of labour. Within the region, such displacement has its precedents. Upheavals north of the Limpopo River have long created a flow of potential workers willing to engage in employment avoided by South Africans (see Chapter 3). However, unlike the 'delayed proletarianisation' of classic regional literature, in which workers relied on a rural base to supplement their wages, many recently displaced Zimbabweans have been the sole source of income for their kin. Some, indeed, with urban backgrounds and total dependence on waged employment, have no rural base at all. Like Polanyi's nineteenth-century workers, displaced Zimbabweans are left entirely at the mercy of capitalists' demands for labour. Indeed, the ad hoc, mobile livelihood strategies pursued by many who come to the border farms as seasonal workers seem perfectly to epitomise a Marxian characterisation of proletarian life, 'free' from prior forms of property or social relations.

For 'respectable' Zimbabweans who find themselves on commercial farms, the experience is one of precipitous decline. In this respect they resemble the country folk whom Polanyi depicts settling in urban industrial settings: 'the moral and cultural catastrophe of the English cottager or copyholder of decent ancestry, who found himself hopelessly sinking

---

examples to broaden the scope of his argument. Indeed, for Polanyi, his story is not merely an English one. He is keen to show parallel processes at work in colonial settings, including in South Africa (ibid.: 164–5). In such settings, 'the disintegration of the cultural environment of the victim is . . . the cause of the degradation' (ibid.: 164). The result, whether in England or in South Africa, is the atomisation of workers' lives, and a disembedding of their labour from social relations.

[6] Because people's productive activities and natural resources are 'fictional commodities' that cannot be reduced to production for sale, and their commoditisation creates impossible conditions for life.

[7] As were earlier patterns of labour migration in Southern Africa. See Chapter 1.

Conclusion 213

in the social and physical slums of some Northwestern factory neighbourhood' (ibid.: 182). As I have shown in this book, what Zimbabwean fragmentation has created is a large population of diverse migrants who, despite their contrasting backgrounds, now represent a large labour reserve on which white farmers and others are able to draw. Zimbabwean displacement appears to have created precisely the disembedding of economic relations that enables uprooted migrants' labour to be bought and sold as easily as possible.

Matters are not quite so simple, however. Understanding the relationship between regional fragmentation and capitalist production requires a closer look at what disembeddedness actually means.

Between Production and Fragmentation: Mediated Paternalism

The discussion above focuses on the circulation of uprooted people as wage-dependent workers. People become detached from wider ties, and are left as mere workers, because they are forced to leave home and follow potential employment. This is disembeddedness as a *precondition* for the creation of workforces. Recall the opening scene of this book: the crowd of transient Zimbabweans who – having travelled long distances, crossed the Limpopo River, and climbed through the border fence – clamoured for work at the Grootplaas gates. In Zimbabwe, as in Polanyi's England, the making of such a labour force is helped along by the fragmenting effects of displacement. Workers' labour is seen as a commodity by the time they arrive for recruitment.

But what about the organisation of work itself? As this book has shown, the impersonalisation of work relations and the emphasis on workers as mere labour are often ways for employers to abrogate responsibility for their employees. Farmers in South Africa increasingly respond to surrounding pressures by treating their workers' labour as simply the object of purchase and sale. They may attempt to negotiate relationships with their workers through the language of markets: a shift from paternalism towards a businesslike, corporate-style managerialism. De-emphasising rooted belonging, including that of workers, distances commercial farming from its past, casting farmers as businesspeople rather than relics of settler colonialism. And doing so enables farmers to negotiate an uncertain future. Moreover, staying afloat increasingly means avoiding undue attachment to particular enterprises. 'There's nothing like sentiment, inheritance, things like that,' one farmer claimed of agriculture in the region. 'What you have to do is be flexible enough to see the opportunities.'

However, as new recruits quickly realise, everyday life at Grootplaas has little to do with employer–employee relations per se. The organisation of work itself often depends on manifold non-contractual relationships and obligations within labour forces (see, for example, Burawoy 1979), which in turn complicate notions of the disembeddedness of labour. Indeed, on the border farms, where employers move towards narrower notions of contract-based work, senior workers may continue to operate according to established arrangements, or even intensify them. They do so to maintain their own status using existing idioms of authority. In doing so, they maintain forms of labour hierarchy that embed work in other aspects of life, creating a mediated paternalism. Workplace authority, forms of seniority outside working hours, and domestic life in the compound are all interrelated. *Mapermanent* place themselves at the centre of enduring webs of vertical dependence. From within the workforce, it can become hard to perceive the flexible pragmatism of white farmers that ultimately sets the terms of farm life.

Breman has observed a 'Great Transformation in the setting of Asia', where 'those who are not fit to work, because they are too young or too old to earn at least their own keep, are discouraged from accompanying members of the household who move off' (2009: 6). The contrast throws the situation on the farms of the Zimbabwean-South African border into sharper relief. The number of dependants; the presence of children; the different forms of labour on the farms, all embedded in the mediated paternalism through which *mapermanent* remain influential – all of this suggests a less stark picture than one where 'labour power, not the social unit of which it is part, is made mobile' (ibid.). The dynamics of formal employment, domestic relationships, and the work of traders and drivers show that 'social units' and 'labour power' cannot be separated here. Mediated paternalism shapes the circulation of people.

Production, in short, need not be based on disembedding workers from social arrangements at all. While *employers* treat employees as commodities, and limit their obligations to contracts, this does not mean that *workers* see one another in the same impersonal manner. Capitalist production, driven by the need to turn a profit while harnessing wider instability, often leads to the decline of paternalist ties at the level of employers/landlords. And yet workforces transform the fragmentation around them and embed workers in new social arrangements – albeit often in limited and provisional ways. On farms like Grootplaas, it is precisely the ways that recruits are incorporated into round-the-clock hierarchies that underpin efficient production and accrue profits to employers.

A Return to Circulation?

This conclusion is not necessarily one that bodes well for the future. As Mollona (2005) notes, the embedding of workplace arrangements in wider social relationships is often a response to intensified exploitation,

Conclusion   215

and one that allows workers to exploit one another. Flexible regimes established by employers can create relations of bonded labour among workers. Social embeddedness is not always of equal benefit to all. The blurring between work and non-work authority in Mollona's case of Sheffield steel workers bears striking resemblance to the material discussed in this book:

> Through their power and authority, embedded in the social hierarchy of the neighbourhood, [the senior workers] coerce the younger workers of the factory and external contractors into production. Outside the shop-floor, the elder workers coerce their children and wives into informal production and control the flow of money and labour. (2005: 544)

Just as pre-capitalist forms of dependency were exploitative, though in different ways to the arrangements of waged labour, so non-contractual ties between workers enable certain powerful figures to benefit at the cost of others. On the border farms, dense webs of relationships mean that work is anything but impersonal and disembedded from social life. Life on the farms is also, however, profoundly unequal.

More broadly, the future of border farming is uncertain – something that the arrival of coal prospectors impressed on workers and their dependants. If employers move, they rupture the dense social ties on which effective work depends. These diverse arrangements and relationships among workers render the end of employment another form of displacement. Shrinking the horizons of workers' plans once again, it leaves them 'free' to sell their labour in a fragmented, mobile population. Even formerly permanent workers may become disconnected from stable social relationships, seeking buyers of their labour power. Seasonal workers will lose a hub of known social relationships, in which working life is embedded.

'Where will we go?' This remains an open question. By 2011, when I returned to visit Grootplaas, coal mining had advanced considerably. Two nearby farmers had sold their land to the mine, and the sounds of machinery and controlled explosions now emanated from the mopaneveld beyond the fields and orchards. At that point, Willem and the other Grootplaas farmers were still considering the possibility of following suit, leasing back the citrus trees to continue production. By 2013, they were no longer pursuing this option. In any case, earlier doubts had not been assuaged – as to whether the water sprays on the mine really kept the coal dust down, and what the long-term effects of the dust would be on crop cultivation along the Limpopo. If necessary, relocation to Mozambique remained an option – but not to the farm they had leased there earlier, which they had abandoned for lack of sufficient water. Amidst the ebb and flow of different strategies, the future has still taken no definite shape.

While it is unclear what will happen to the border farms, the dangers invite a fittingly Polanyian gloss, as land itself is further 'disembedded' from the complex web of relationships in which people live. To be sure, land is already privately owned in this setting, and the farmers' monopoly of ownership shapes workers' lives profoundly. But now, the effect of instability – partly the result of the vagaries of land reform – is precisely to shift farmers' views of their enterprises more aggressively towards a model of transactive business. The effect, in Polanyi's words, is 'to separate land from man and to organise society in such a way as to satisfy the requirements of a real-estate market' (1944: 187). For workers, this means that they not only face a future as part of a regional labour reserve (even more so than previously), but also have to contend with employers themselves moving, as locations are assessed by reference to their profitability and security for white farmers. The 'disembedding' of labour and land, and their consolidation as commodities for purchase, are inseparable in practice.

The temporality of this uncertainty is itself complex. There is a sharp disjuncture between the embeddedness of everyday existence on the farms, and the instability of the future. Such instability has been seen as characteristic of flexible capitalism and its ephemerality (Harvey 1990: 171). In this view, people's stances towards the future shift in response, 'abandon[ing] commitments and loyalties without regret' (Bauman 2007: 4), even taking on character traits that reflect the organisation of production – 'the capacity to let go of one's past, the confidence to accept fragmentation' (Sennett 1998: 63). The ground constantly shifts; people must learn to shift with it.

But on the border, workers strive for a provisional permanence (Bolt 2013). They have little choice but to see their immediate rootedness at the farm as real, and the more distant future as the realm of conjecture. They continue to adapt accommodation, develop their homes, and invest in relationships in the compound. And indeed farmers, for now, continue to commit to their projects on the farms, driven by the well-worn logic of capitalist expansion. This, even while the more enterprising spread their risk.

Rather than producing ephemerality per se, flexibility and uncertainty create a temporal limbo. At Grootplaas, workers and farmers plan in the short term as though the farm will remain operating forever. Workers lack other options. Even Michael, who during my fieldwork was sure that his qualifications would make him a valuable asset elsewhere in South Africa if he left the farm, had by 2011 come to realise this. Willem had sidelined him after discovering his attempts to embezzle money intended for work permits. This had led not only to narrower responsibilities, but also to insecure pay, a much smaller room, and broader marginalisation within the workforce and the compound. Despite these humiliations – and despite the fact that he had finally finished building his rural home[8] – he hung on, hoping to be brought back into the fold. And indeed, the number of highly qualified Zimbabweans seeking work makes any better-paid or higher-status employment doubtful.

Farmers' enterprises, meanwhile, only survive through continual investment and competitive expansion. The difference is that farmers are planning at a larger scale, in the process limiting their attachment to estates, and creating the conditions for long-term impermanence. Even so, their enterprises remain rooted in places, characterised by enduring social, ethical, and temporal arrangements. Everyday life occurs at a remove from impending change. It is hard for workers to see what might happen next – where they will lie on the sliding scale of settlement and transience – on the banks of the Limpopo.

---

[8] He had begun accumulating building materials by 2006 but, during a trip we made together to the site in rural Beitbridge District in 2007, he discovered that his carefully amassed corrugated roofing and window frames had been incorporated into his uncle's own construction project. When I returned to the farm to visit in 2011, he showed me a photograph of a small but recently completed concrete house in the corner of a rectangular compound.

References

Abraham, I. and W. van Schendel. 2005. 'Introduction: the making of illicitness', in I. Abraham and W. van Schendel (eds), *Illicit Flows and Criminal Things: states, borders, and the other side of globalisation*. Bloomington, IN: Indiana University Press.

Addison, L. 2006. 'Frontier Farm Labour: a study of neoliberal restructuring and Zimbabwean migrant farm workers in Limpopo Province, South Africa'. MA dissertation, Carleton University, Ottawa.

—— 2013. 'Factions in the field: social divisions and gendered survival strategies on a South African border farm', in B. Derman and R. Kaarhus (eds), *In the Shadow of a Conflict: crisis in Zimbabwe and its effects in Mozambique, South Africa and Zambia*. Harare: Weaver Press.

Alden, C. and W. Anseeuw. 2009. *Land, Liberation and Compromise in Southern Africa*. Basingstoke: Palgrave Macmillan.

Alexander, J. 2003. '"Squatters", veterans and the state in Zimbabwe', in A. Hammar, B. Raftopoulos, and S. Jensen (eds), *Zimbabwe's Unfinished Business: rethinking land, state and nation in the context of crisis*. Harare: Weaver Press.

——, J. McGregor, and T. Ranger. 2000. *Violence and Memory: one hundred years in the 'dark forests' of Matabeleland*. Oxford: James Currey.

Alvarez Jr., R. R. 1995. 'The Mexican-US border: the making of an anthropology of borderlands', *Annual Review of Anthropology* 24: 447–70.

Andersson, J. A. 2006. 'Informal moves, informal markets: international migrants and traders from Mzimba District, Malawi', *African Affairs* 105 (420): 375–97.

Arrighi, G. 1970. 'Labour supplies in historical perspective: a study of the proletarianisation of the African peasantry in Rhodesia, *Journal of Development Studies* 6 (3): 197–234.

Augé, M. 1995. *Non-Places: introduction to an anthropology of supermodernity*, translated by John Howe. London: Verso.

Bakewell, O. 2008. '"Keeping them in their place": the ambivalent relationship between development and migration in Africa', *Third World Quarterly* 29 (7): 1341–58.

Ballard, R. 2003. 'A case of capital-rich under-development: the paradoxical consequences of successful transnational entrepreneurship from Mirpur', *Contributions to Indian Sociology*, N.S. 37 (1 and 2): 49–81.

Bank, L. 1999. 'Men with cookers: transformations in migrant culture, domesticity and identity in Duncan Village, East London', *Journal of Southern African Studies* 25 (3): 393–416.

Barchiesi, F. 2011. *Precarious Liberation: workers, the state, and contested social citizenship in postapartheid South Africa*. Albany, NY: State University of New York Press.

Barnes, T. 1999. *'We Women Worked So Hard': gender, urbanisation and social reproduction in colonial Harare, 1930–1956*. Oxford: James Currey.

Bauman, Z. 2007. *Liquid Times: living in an age of uncertainty*. Cambridge: Polity.

Bayart, J.-F. 2000. 'Africa in the world: a history of extraversion', *African Affairs* 99 (395): 217–67.

Beinart, W. 2001. *Twentieth Century South Africa*. Oxford: Oxford University Press.

Beinart, W., P. Delius, and S. Trapido. 1986. 'Introduction', in W. Beinart, P. Delius, and S. Trapido (eds), *Putting a Plough to the Ground: accumulation and dispossession in rural South Africa 1850–1930*. Johannesburg: Ravan Press.

Beremauro, R. 2013. 'Living Between Compassion and Domination? An Ethnographic Study of Institutions, Interventions and the Everyday Practices of Poor Black Zimbabwean Migrants in South Africa'. PhD thesis, University of the Witwatersrand, Johannesburg.

Bernstein, H. 1996. 'South Africa's agrarian question: extreme and exceptional?', *Journal of Peasant Studies* 23 (2): 1–52.

2004. 'The boys from Bothaville, or the rise and fall of King Maize: a South African story', *Journal of Agrarian Change* 4 (4): 492–508.

2007. 'Agrarian questions of capital and labour: some theory about land reform (and a periodisation)', in L. Ntsebeza and R. Hall (eds), *The Land Question in South Africa: the challenge of transformation and redistribution*. Cape Town: HSRC Press.

Beynon, H. 1973. *Working for Ford*. Harmondsworth: Penguin.

Bhebe, N. and T. Ranger (eds). 1995a. *Soldiers in Zimbabwe's Liberation War*. London: James Currey.

1995b. 'Volume introduction: soldiers in Zimbabwe's Liberation War', in N. Bhebe and T. Ranger (eds), *Soldiers in Zimbabwe's Liberation War*. London: James Currey.

(eds). 1996. *Society in Zimbabwe's Liberation War*. Oxford: James Currey.

Blum, J. 2000. 'Degradation without deskilling: twenty-five years in the San Francisco shipyards', in M. Burawoy (ed.), *Global Ethnography: forces, connections, and imaginations in a postmodern world*. Berkeley, CA: University of California Press.

Boeyens, J. C. A. 1994. '"Black ivory": the indenture system and slavery in Zoutpansberg, 1848–1869', in E. A. Eldridge and F. Morton (eds), *Slavery in South Africa: captive labour on the Dutch frontier*. Oxford: Westview Press.

Bohannan, P. 1955. 'Some principals of exchange and investment among the Tiv', *American Anthropologist* 57 (1): 60–70.

Bolt, M. 2010. 'Camaraderie and its discontents: class consciousness, ethnicity and divergent masculinities among Zimbabwean migrant farmworkers in South Africa', *Journal of Southern African Studies* 36 (2): 377–93.

2012. 'Waged entrepreneurs, policed informality: work, the regulation of space and the economy of the Zimbabwean-South African border', *Africa* 82 (1): 111–30.

2013. 'Producing permanence: employment, domesticity and the flexible future on a South African border farm', *Economy and Society* 42 (2): 197–225.

2014. 'The sociality of the wage: money rhythms, wealth circulation and the problem with cash on the Zimbabwean-South African border', *Journal of the Royal Anthropological Institute* N.S. 20 (1): 113–30.

Bond, P. 2000. *Elite Transition: from apartheid to neoliberalism in South Africa*. London: Pluto Press.

Bonner, P. 2003a. 'The hunting frontier in the west', in P. Bonner and E. J. Carruthers, *The Recent History of the Mapungubwe Area*. Mapungubwe Cultural Heritage Resources Survey. Commissioned by the South African Department of Environmental Affairs and Tourism.

2003b. 'Reconsolidating white power in the far north 1877–1899', in P. Bonner and E. J. Carruthers, *The Recent History of the Mapungubwe Area*. Mapungubwe Cultural Heritage Resources Survey. Commissioned by the South African Department of Environmental Affairs and Tourism.

2003c. 'Mining in the Messina/Mapungubwe area', in P. Bonner and E. J. Carruthers, *The Recent History of the Mapungubwe Area*. Mapungubwe Cultural Heritage Resources Survey. Commissioned by the South African Department of Environmental Affairs and Tourism.

2003d. 'Introduction', in P. Bonner and E. J. Carruthers, *The Recent History of the Mapungubwe Area*. Mapungubwe Cultural Heritage Resources Survey. Commissioned by the South African Dept of Environmental Affairs and Tourism.

P. Delius, and D. Posel. 1993. 'The shaping of apartheid: contradiction, continuity and popular struggle', in P. Bonner, P. Delius, and D. Posel (eds), *Apartheid's Genesis, 1935–1962*. Johannesburg: Ravan Press.

Bourdillon, M. F. C. 1987. 'Guns and rain: taking structural analysis too far?', *Africa* 57 (2): 263–73.

Bozzoli, B. 1991. 'The meaning of informal work: some women's stories', in E. Preston-Whyte and C. Rogerson (eds), *South Africa's Informal Economy*. Oxford: Oxford University Press.

Bozzoli, B. with M. Nkotsoe. 1991. *Women of Phokeng: consciousness, life strategy and migrancy in South Africa, 1900–1983*. London: James Currey.

Bradford, H. 1991. 'Highways, byways and culs-de-sac: the transition to agrarian capitalism in revisionist South African history', in J. Brown (ed.), *History from South Africa: alternative visions and practices*. Philadelphia, PA: Temple University Press.

1993. 'Getting away with murder: "mealie kings", the state and foreigners in the Eastern Transvaal, c. 1918–1950', in P. Bonner, P. Delius and D. Posel (eds), *Apartheid's Genesis, 1935–1962*. Johannesburg: Ravan Press.

Braun, L. F. 2013. 'The returns of the king: the case of Mphephu and Western Venda, 1899–1904', *Journal of Southern African Studies* 39 (2): 271–91.

Breman, J. 2009. 'The Great Transformation in the setting of Asia'. Address delivered on the occasion of the award of the degree *Doctor Honoris Causa* on the 57th Anniversary of the International Institute of Social Studies, The Hague, The Netherlands, 29 October.

Burawoy, M. 1979. *Manufacturing Consent: changes in the labour process under monopoly capitalism*. Chicago, IL: Chicago University Press.

Burke, T. 1996. *Lifebuoy Men, Lux Women: commodification, consumption and cleanliness in modern Zimbabwe*. Durham, NC: Duke University Press.

Campbell, C. 2001. '"Going underground and going after women": masculinity and HIV transmission amongst black workers on the gold mines', in R. Morrell (ed.), *Changing Men in Southern Africa*. London: Zed Books.

Castells, M. and J. Henderson. 1987. 'Introduction: techno-economic restructuring, socio-political processes and spatial transformation: a global perspective', in J. Henderson and M. Castells (eds), *Global Restructuring and Territorial Development*. London: Sage.

Castree, N. 2009. 'The spatio-temporality of capitalism', *Time and Society* 18 (1): 26–61.

Chabal, P. and J.-P. Daloz. 1999. *Africa Works: disorder as a political instrument*. Oxford: James Currey.

Cliffe, L., J. Alexander, B. Cousins, and R. Gaidzanwa. 2011. 'An overview of Fast Track Land Reform in Zimbabwe: editorial introduction', *Journal of Peasant Studies* 38 (5): 907–38.

Coetzee, J. M. 2004 [1976]. *In the Heart of the Country*. London: Vintage.

Collinson, D. L. 1988. 'Engineering humour: masculinity, joking and conflict in shopfloor relations', *Organization Studies* 9 (2): 181–99.

Comaroff, J. 1985. *Body of Power, Spirit of Resistance: the culture and history of a South African people*. Chicago, IL: University of Chicago Press.

and J. L. Comaroff. 1987. 'The madman and the migrant: work and labour in the historical consciousness of a South African people', *American Ethnologist* 14 (2): 191–209.

1989. 'The colonisation of consciousness in South Africa'. *Economy and Society* 18 (3): 267–96.

1990. 'Goodly beasts, beastly goods: cattle and commodities in a South African context'. *American Ethnologist* 17 (2): 195–216.

1999. 'Occult economies and the violence of abstraction: notes from the South African postcolony', *American Ethnologist* 26 (2): 279–303.

2000. 'Millennial capitalism: first thoughts on a second coming', *Public Culture* 12 (2): 291–343.

Connell, R. W. 2005. *Masculinities (2nd Edition)*. Berkeley, CA: University of California Press.

Cooper, F. 2005. *Colonialism in Question: theory, knowledge, history*. London: University of California Press.

Coplan, D. B. 1987. 'Eloquent knowledge: Lesotho migrants' songs and the anthropology of experience', *American Ethnologist* 14 (3): 413–33.

1991. 'Fictions that save: migrants' performance and Basotho national culture', *Cultural Anthropology* 6 (2): 164–92.

2001. 'A river runs through it: the meaning of the Lesotho-Free State border', *African Affairs* 100 (398): 81–116.

Crapanzano, V. 1985. *Waiting: the Whites of South Africa*. New York, NY: Random House.

Crush, J. 2000. 'Introduction: making hay with foreign farmworkers', in J. Crush (ed.), *Borderline Farming: foreign migrants in South African commercial agriculture*. Southern African Migration Project Migration Policy Series No. 16. Cape Town: Idasa.

Dangarembga, T. 2004. *Nervous Conditions*. Banbury: Ayebia Clarke.

Day, S., E. Papataxiarchis, and M. Stewart. 1998. 'Consider the lilies of the field', in S. Day, E. Papataxiarchis, and M. Stewart (eds), *Lilies of the Field: marginal people who live for the moment*. Boulder, CO: Westview Press.

Delius, P. 1980. 'Migrant labour and the Pedi, 1840–80', in S. Marks and A. Atmore (eds), *Economy and Society in Pre-Industrial South Africa*. London: Longman.

and S. Trapido. 1982. 'Inboekselings and oorlams: the creation and transformation of a servile class', *Journal of Southern African Studies* 8 (2): 214–42.

Derman, B. 2013. 'Governing the South African/Zimbabwean border: immigration, criminalisation and human rights', in B. Derman and R. Kaarhus (eds), *In the Shadow of a Conflict: crisis in Zimbabwe and its effects in Mozambique, South Africa and Zambia.* Harare: Weaver Press.

and R. Kaarhus 2013. 'Introduction: crisis in Zimbabwe and its regional effects', in B. Derman and R. Kaarhus (eds), *In the Shadow of a Conflict: crisis in Zimbabwe and its effects in Mozambique, South Africa and Zambia.* Harare: Weaver Press.

Doctors Without Borders. 2009. 'Beyond cholera: Zimbabwe's worsening crisis', 17 February. <http://doctorswithoutborders.org/publications/article.cfm?id=3408&cat=special-report> (accessed 28 August 2013).

Donnan, H. and T. M. Wilson. 1999. *Borders: frontiers of identity, nation and state.* Oxford: Berg.

Dorman, S. R. 2003. 'NGOs and the constitutional debate in Zimbabwe: from inclusion to exclusion', *Journal of Southern African Studies* 29 (4): 845–63.

du Toit, A. 1993. 'The micro-politics of paternalism: the discourses of management and resistance on South African fruit and wine farms', *Journal of Southern African Studies* 19 (2): 314–36.

Elder, G. S. 2003. *Hostels, Sexuality, and the Apartheid Legacy: malevolent geographies.* Athens, OH: Ohio University Press.

Englund, H. 2002. *From War to Peace on the Mozambique-Malawi Borderland.* Edinburgh: International African Institute and Edinburgh University Press.

Epprecht, M. 1998. 'The "unsaying" of indigenous homosexualities in Zimbabwe: mapping a blindspot in an African masculinity', *Journal of Southern African Studies* 24 (4): 631–51.

Epstein, A. L. 1958. *Politics in an Urban African Community.* Manchester: Manchester University Press.

Erlmann, V. 1992. '"The past is far and the future is far": power and performance among Zulu migrant workers', *American Ethnologist* 19 (4): 688–709.

Ewert, J. and J. Hamman. 1999. 'Why paternalism survives: globalisation, democratization and labour on South African wine farms', *Sociologia Ruralis* 39 (2): 202–21.

Ewert, J. and A. du Toit. 2005. 'A deepening divide in the countryside: restructuring and rural livelihoods in the South African wine industry', *Journal of Southern African Studies* 31 (2): 315–32.

Ferguson, J. 1985. 'The bovine mystique: power, property and livestock in rural Lesotho', *Man* N.S. 20 (4): 647–74.

1992. 'The cultural topography of wealth: commodity paths and the structure of property in rural Lesotho', *American Anthropologist* 94 (1): 55–73.

1999. *Expectations of Modernity: myths and meanings of urban life on the Zambian Copperbelt.* Berkeley, CA: University of California Press.

2007. 'Formalities of poverty: thinking about social assistance in neoliberal South Africa', *African Studies Review* 50(2): 71–86.

2013. 'Declarations of dependence: labor, personhood, and welfare in southern Africa', *Journal of the Royal Anthropological Institute* 19 (2): 223–42.

Fernandez-Kelly, P. 2006. 'Introduction', in P. Fernandez-Kelly and J. Shefner (eds), *Out of the Shadows: political action and the informal economy in Latin America.* University Park, PA: Pennsylvania State University Press.

Feuchtwang, S. 2004. 'Theorising place', in S. Feuchtwang (ed.), *Making Place: state projects, globalisation and local responses in China*. London: UCL Press.
Finnström, S. 2008. *Living with Bad Surroundings: war, history, and everyday moments in northern Uganda*. Durham, NC: Duke University Press.
Flynn, D. K. 1997. '"We are the border": identity, exchange and the state along the Bènin-Nigeria border, *American Ethnologist* 24 (2): 311–30.
Fuller, A. 2002. *Don't Let's Go to the Dogs Tonight: an African childhood*. New York, NY: Random House.
Fultz, E. with B. Pieris. 1997. 'The social protection of migrant workers in South Africa'. ILO/SAMAT Policy Paper No. 3. <http://www.ilo.org/public/french/region/afpro/pretoria/papers/1997/polpap3/index.htm> (accessed 9 April 2014).
Gamburd, M. 2004. 'Money that burns like oil: a Sri Lankan cultural logic of morality and agency', *Ethnology* 43 (2): 167–84.
GAPWUZ (General Agricultural and Plantation Workers' Union of Zimbabwe). 2010. *If Something is Wrong: the invisible suffering of farmworkers due to 'land reform'*. Harare: Weaver Press.
Gardner, K. and F. Osella. 2003. 'Migration, modernity and social transformation in South Asia: an overview', *Contributions to Indian Sociology* 37 (1 and 2): v–xxviii.
Geschiere, P. and F. Nyamnjoh. 2000. 'Capitalism and autochthony: the seesaw of mobility and belonging', *Public Culture* 12 (2): 423–52.
Gibbon, P. 1995. 'Introduction: structural adjustment and the working poor in Zimbabwe' in P. Gibbon (ed.), *Structural Adjustment and the Working Poor in Zimbabwe: studies on labour, women informal sector workers and health*. Uppsala: Nordiska Afrikainstitutet.
Gilliomee, H. 2003. *The Afrikaners: biography of a people*. Cape Town: Tafelberg.
Ginsberg, R. 1996. '"Now I stay in a house": renovating the matchbox in apartheid-era Soweto', *African Studies* 55 (2): 127–39.
Gluckman, M. 1961. 'Anthropological problems arising from the African Industrial Revolution', in A. Southall (ed.), *Social Change in Modern Africa: studies presented and discussed at the First International African Seminar*. Oxford: Oxford University Press.
Godwin, P. and I. Hancock. 1993. *Rhodesians Never Die: the impact of war and political change on White Rhodesia, c.1970–1980*. Oxford: Oxford University Press.
Gordon, R. J. 1977. *Mines, Masters and Migrants: life in a Namibian compound*. Johannesburg: Ravan Press.
Guardian. 2008. 'Zimbabwe inflation passes 100,000%, officials say', 22 February. <http://www.theguardian.com/world/2008/feb/22/zimbabwe> (accessed 28 August 2013).
Guardian. 2015. 'South African church closes doors to refugees after sheltering 30,000', 1 January. <http://www.theguardian.com/world/2015/jan/01/south-african-church-refugees-johannesburg> (accessed 3 July 2015).
Gudeman, S. 2010. 'A cosmopolitan anthropology?' in D. James, E. Plaice, and C. Toren (eds), *Culture Wars: context, models, and anthropologists' accounts*. Oxford: Berghahn.
Guy, J. and M. Thabane, 1987. 'The Ma-Rashea: a participant's perspective', in B. Bozzoli (ed.), *Class, Community and Conflict: South African perspectives*. Johannesburg: Ravan Press.

1991. 'Basotho miners, ethnicity and workers' strategies', in I. Brandell (ed.), *Workers in Third-World Industrialization*. London: Macmillan.
Guyer, J. I. 2004. *Marginal Gains: monetary transactions in Atlantic Africa*. Chicago, IL and London: University of Chicago Press.
Hall, R. 2003. *Farm Tenure*. Evaluating Land and Agrarian Reform in South Africa Series No. 3. Cape Town: Programme for Land and Agrarian Studies.
2007. 'Transforming rural South Africa? Taking stock of land reform', in L. Ntsebeza and R. Hall (eds), *The Land Question in South Africa: the challenge of transformation and redistribution*. Cape Town: HSRC Press.
2013. 'Hierarchies, violence, gender: narratives from Zimbabwean migrants on South African farms', in B. Derman and R. Kaarhus (eds), *In the Shadow of a Conflict: crisis in Zimbabwe and its effects in Mozambique, South Africa and Zambia*. Harare: Weaver Press.
Hammar, A. 2003. 'The making and unma(s)king of local government in Zimbabwe', in A. Hammar, B. Raftopoulos, and S. Jensen (eds), *Zimbabwe's Unfinished Business: rethinking land, state and nation in the context of crisis*. Harare: Weaver Press.
2010. 'Ambivalent mobilities: Zimbabwean commercial farmers in Mozambique', *Journal of Southern African Studies* 36 (2): 395–416.
2014. 'The paradoxes of class: crisis, displacement and repositioning in post-2000 Zimbabwe, in A. Hammar (ed.), *Displacement Economies in Africa: paradoxes of crisis and creativity*. London: Zed Books.
J. McGregor, and L. Landau. 2010. 'Introduction. Displacing Zimbabwe: crisis and construction in southern Africa', *Journal of Southern African Studies* 36 (2): 263–83.
and B. Raftopoulos. 2003. 'Zimbabwe's unfinished business: rethinking land, state and nation' in A. Hammar, B. Raftopoulos, and S. Jensen (eds), *Zimbabwe's Unfinished Business: rethinking land, state and nation in the context of crisis*. Harare: Weaver Press.
Hammond-Tooke, W. D. 1997. *Imperfect Interpreters: South Africa's anthropologists 1920–1990*. Johannesburg: Witwatersrand University Press.
Hart, G. 2002. *Disabling Globalization: places of power in post-apartheid South Africa*. Berkeley, CA: University of California Press
Hart, K. 1973. 'Informal income opportunities and urban employment in Ghana', *Journal of Modern African Studies* 11 (1): 61–89.
Harvey, D. 1990. *The Condition of Postmodernity*. Oxford: Blackwell.
HRW (Human Rights Watch). 2006a. *Unprotected Migrants: Zimbabweans in South Africa's Limpopo Province*. New York, NY: Human Rights Watch, Volume 18, No. 6 (A).
2006b. 'Zimbabwe: events of 2005'. <http://www.hrw.org/en/world-report-2006/zimbabwe> (accessed 28 August 2013).
Hughes, D. M. 2010. *Whiteness in Zimbabwe: race, landscape and the problem of belonging*. New York, NY: Palgrave Macmillan.
Hull, E. and D. James. 2012. 'Introduction: popular economies in South Africa', *Africa* 82 (1): 1–20.
ILO (International Labour Office). 1998. 'Labour migration to South Africa in the 1990s'. Policy Paper Series No.4. Harare: International Labour Office.
IRIN (Integrated Regional Information Networks). 2008. 'Zimbabwe: tortured, raped and forgotten', 23 September. <http://www.irinnews.org/Report.aspx?ReportId=80544> (accessed 18 November 2010).

2013. 'Deported from South Africa, Zimbabweans struggle at home', 6 June. <http://www.irinnews.org/report/98174/deported-from-south-africa-zimbabweans-struggle-at-home> (accessed 28 August 2013).

James, D. 1990. 'A question of ethnicity: Ndundza Ndebele in a Lebowa village', *Journal of Southern African Studies* 16 (1): 33–54.

1999. *Songs of the Women Migrants: performance and identity in South Africa*. Edinburgh: Edinburgh University Press.

2001. 'Land for the landless: conflicting images of rural and urban in South Africa's land reform programme', *Journal of Contemporary African Studies* 19 (1): 93–109.

2007. *Gaining Ground? 'Rights' and 'property' in South African land reform*. Johannesburg: Wits University Press.

2009. 'Burial sites, informal rights and lost kingdoms: contesting land claims in Mpumalanga, South Africa', *Africa* 79 (2): 228–51.

n.d. '"Music of origin": class, social category and the performers and audience of *Kiba*, a South African migrant genre', unpublished paper.

and E. Hull (eds). 2012. *Popular Economies in South Africa*. Special Issue of *Africa* 82 (1).

Al Jazeera. 2015. 'Church housing refugees raided by South African police', 8 May. <http://www/aljazeera.com/news/2015/05/church-housing-refugees-raided-south-africa-police-150508214742974.html> (accessed 3 July 2015).

Jeeves, A. H. 1983. 'Over-reach: the South African gold mines and the struggle for the labour of Zambesia, 1890–1920', *Canadian Journal of African Studies* 17 (3): 393–412.

Jensen, S. and L. Buur, 2007. 'The nationalist imperative: South Africanisation, regional integration and mobile livelihoods', in L. Buur, S. Jensen, and F. Stepputat (eds), *The Security-Development Nexus: expressions of sovereignty and securitisation in southern Africa*. Uppsala: Nordiska Afrikainstitutet.

Jing, J. 2003. 'A return to the homeland movement in northwest China', in C. Stafford (ed.), *Living with Separation in China: anthropological accounts*. London: Routledge Curzon.

Johnston, D. 2007. 'Who needs immigrant farm workers? A South African case study', *Journal of Agrarian Change* 7 (4): 494–525.

Jones, J. L. 2010. '"Nothing is straight in Zimbabwe": the rise of the kukiya-kiya economy 2000–2008', *Journal of Southern African Studies* 36 (2): 285–99.

2014. '"No move to make": the Zimbabwe crisis, displacement-in-place and the erosion of "proper places"', in A. Hammar (ed.), *Displacement Economies in Africa: paradoxes of crisis and creativity*. London: Zed Books.

Koelble, T. A. and E. LiPuma. 2005. 'Traditional leaders and democracy: cultural politics in the age of globalisation' in S. L. Robins (ed.), *Limits to Liberation after Apartheid: citizenship, governance and culture*. Oxford: James Currey.

Kosmin, B. 1977. 'The Inyoka tobacco industry of the Shangwe people: the displacement of a pre-colonial economy in Southern Rhodesia 1898–1938, in R. Palmer and Parsons, N. (eds) *The Roots of Rural Poverty in Central and Southern Africa*. London: Heinemann.

Kriger, N. 1992. *Zimbabwe's Guerrilla War: peasant voices*. Cambridge: Cambridge University Press.

2010. 'The politics of legal status for Zimbabweans in South Africa', in J. McGregor and R. Primorac (eds), *Zimbabwe's New Diaspora: displacement and the cultural politics of survival*. Oxford: Berghahn.

Lan, D. 1985. *Guns and Rain: guerrillas and spirit mediums in Zimbabwe*. Oxford: James Currey.

Lee, R. 2005. 'Reconstructing "home" in apartheid Cape Town: African women and the process of settlement', *Journal of Southern African Studies* 31 (3): 611–30.

Lessing, D. 2007 (1950). *The Grass Is Singing*. London: Harper Perennial.

Lincoln, D. and C. Maririke. 2000. 'Southward migrants in the far north: Zimbabwean farmworkers in Northern Province', in J. Crush (ed.), *Borderline Farming: foreign migrants in South African commercial agriculture*. Southern African Migration Project Migration Policy Series No. 16. Cape Town: Idasa.

Lindquist, J. 2009. *The Anxieties of Mobility: migration and tourism in the Indonesian borderlands*. Honolulu, HI: University of Hawai'i Press.

Lodge, T. 2003. *Politics in South Africa: from Mandela to Mbeki*. Oxford: James Currey.

Loizos, P. 2008. *Iron in the Soul: displacement, livelihood and health in Cyprus*. Oxford: Berghahn.

MacDonald, A. 2012. 'Colonial Trespassers in the Making of South Africa's International Borders 1900 to c.1950'. PhD thesis, University of Cambridge.

MacGaffey, J. 1991. *The Real Economy of Zaire: the contribution of smuggling and other unofficial activities to national wealth*. London: James Currey.

—— and R. Bazenguissa-Ganga. 2000. *Congo-Paris: transnational traders on the margins of the law*. Oxford: James Currey.

Mail and Guardian. 2008. 'Zimbabwe licenses shops to accept foreign currency', 26 September. <http://mg.co.za/article/2008-09-26-zimbabwe-licenses-shops-to-accept-foreign-currency> (accessed 28 August 2013).

Malkki, L. 1992. 'National geographic: the rooting of peoples and the territorialization of national identity among scholars and refugees', *Cultural Anthropology* 7 (1): 24–44.

Mamadi, M. F. 1942. 'The copper miners of Musina', in N.J. van Warmelo (ed.), *The Copper Miners of Musina and the Early History of the Zoutpansberg*, Union of South Africa Ethnological Publications Vol. VIII. Pretoria: Union of South Africa Government Printer.

Maphosa, F. 2005. 'The impact of remittances from Zimbabweans working in South Africa on rural livelihoods in the southern districts of Zimbabwe', *Forced Migration Working Paper Series* 14, University of the Witwatersrand, Johannesburg.

Mate, R. 2005. *Making Ends Meet at the Margins? Grappling with economic crisis and belonging in Beitbridge town, Zimbabwe*. Dakar: CODESRIA.

Matondi, P. B. 2012. *Zimbabwe's Fast Track Land Reform*. London: Zed Books.

Mayer, P. 1980. 'The origin and decline of two rural resistance ideologies', in P. Mayer (ed.), *Black Villagers in an Industrial Society*. Oxford: Oxford University Press.

Mazzucato V., M. Kabki, and L. Smith. 2006. 'Transnational migration and the economy of funerals: changing practices in Ghana'. *Development and Change* 37 (5): pp.1047–72.

Mbembe, A. and J. Roitman. 1995. 'Figures of the subject in times of crisis', *Public Culture* 7 (2): 323–52.
McAllister, P. 1980. 'Work, homestead and the shades: the ritual interpretation of labour migration among the Gcaleka', in P. Mayer (ed.), *Black Villagers in an Industrial Society*. Oxford: Oxford University Press.
  1985. 'Beasts to beer pots: migrant labour and ritual change in Willowvale District, Transkei', *African Studies* 44 (2): 121–35.
  1991. 'Using ritual to resist domination in the Transkei', in A. D. Spiegel and P. A. McAllister (eds), *Tradition and Transition in Southern Africa: African Studies Fiftieth Anniversary Volume* 50 (1 and 2): 129–44.
McGregor, J. 2008. 'Abject spaces, transnational calculations: Zimbabweans in Britain navigating work, class and the law', *Transactions of the Institute of British Geographers* 33 (4): 466–82.
  2010. 'Introduction: the making of Zimbabwe's new diaspora', in J. McGregor and R. Primorac (eds), *Zimbabwe's New Diaspora: displacement and the cultural politics of survival*. Oxford: Berghahn.
McKay, D. 2003. 'Cultivating new local futures: remittance economies and land-use patterns in Ifugao, Philippines', *Journal of Southeast Asian Studies* 34 (2): 285–306.
McNamara, J. K. 1978. 'Social Life, Ethnicity and Conflict in a Gold Mine Hostel'. MA dissertation, University of the Witwatersrand, Johannesburg.
  1985. 'Black Worker Conflicts on South African Gold Mines: 1973–1982'. PhD thesis, University of the Witwatersrand, Johannesburg.
McNeill, F. G. 2009. '"Condoms cause AIDS": poison, prevention and denial in Venda, South Africa', *African Affairs* 108 (432): 353–70.
Meagher, K. 2010. *Identity Economics: social networks and the informal economy in Nigeria*. Woodbridge: James Currey.
Meillassoux, C. 1981. *Maidens, Meals and Money: capitalism and the domestic economy*. Cambridge: Cambridge University Press.
Messina (Transvaal) Development Company. 1954a. *Copper Cavalcade: 50 years of the Messina Copper Mine*. Messina (Transvaal) Development Company.
  1954b. *A Brief History*. Messina (Transvaal) Development Company.
Messina Publicity Association. c.1961. *Messina: the most northern town in the Republic*. Messina Publicity Association.
Mills, J. A. G. 1952. *Messina, Northern Transvaal: a short history*. Messina Mail-Coach Organising Committee of the Van Riebeeck Festival.
Mollona, M. 2005. 'Factory, family and neighbourhood: the political economy of informal labour in Sheffield', *Journal of the Royal Anthropological Institute* 11 (3): 527–48.
Moodie, T. D. 1980. 'The formal and informal structure of a South African gold mine', *Human Relations* 33 (8): 555–74.
  1983. 'Mine culture and miners' identity on the South African gold mines' in B. Bozzoli (ed.), *Town and Country in the Transvaal: capitalist penetration and popular response*. Johannesburg: Ravan Press.
Moodie, T. D. with V. Ndatshe. 1994. *Going For Gold: mines, men and migration*. Berkeley, CA: University of California Press.
Morreira, S. 2010. 'Seeking solidarity: Zimbabwean undocumented migrants in Cape Town, 2007', *Journal of Southern African Studies* 36 (2): 433–48.
Morrell, R. 2001. 'The times of change: men and masculinity in South Africa', in R. Morrell (ed.), *Changing Men in Southern Africa*. London: Zed Books.

Motenda, M. M. 1942. 'History of the western Venda and of the Lemba', in N. J. van Warmelo (ed.), *The Copper Miners of Musina and the Early History of the Zoutpansberg*, Union of South Africa Ethnological Publications Vol. VIII. Pretoria: Union of South Africa Government Printer.

Mtisi, J., M. Nyakudya, and T. Barnes. 2009. 'War in Rhodesia, 1965–1980', in B. Raftopoulos and A. S. Mlambo (eds), *Becoming Zimbabwe: a history from the pre-colonial period to 2008*. Harare: Weaver Press.

Murray, C. 1981. *Families Divided: the impact of migrant labour in Lesotho*. Cambridge: Cambridge University Press.

Murray, M. 1995. 'Blackbirding at "Crooks' Corner": illicit labour recruitment in the Northeastern Transvaal 1910–1940', *Journal of Southern African Studies* 21 (3): 373–97.

Muzondidya, J. 2009. 'From buoyancy to crisis, 1980–1997', in B. Raftopoulos and A. S. Mlambo (eds), *Becoming Zimbabwe: a history from the pre-colonial period to 2008*. Harare: Weaver Press.

Muzvidziwa, V. 2001. 'Zimbabwe's cross-border women traders: multiple identities and responses to new challenges'. *Journal of Contemporary African Studies* 19 (1): 67–80.

Ndlovu-Gatsheni, S. J. 2009. *Do 'Zimbabweans' Exist?: trajectories of nationalism, national identity formation and crisis in a postcolonial state*. Oxford: Peter Lang.

Newman, K. S. 1988. *Falling from Grace: the experience of downward mobility in the American middle class*. London: The Free Press (Collier Macmillan).

Norman, W. O. 2005. 'Living on the Frontline: politics, migration and transfrontier conservation in the Mozambican villages of the Mozambique-South Africa borderland.' PhD thesis, University of London (LSE), London.

Nyamnjoh, F. B. 2006. *Insiders and Outsiders: citizenship and xenophobia in contemporary Southern Africa*. London: Zed Books.

Oomen, B. 2005. *Chiefs in South Africa: law, power and culture in the post-apartheid era*. Oxford: James Currey.

Orner, P. and A. Holmes (eds). 2011. *Don't Listen to What I'm About to Say: voices of Zimbabwe*. Jeppestown: Jonathan Ball.

Ortiz, S. 2002. 'Laboring in the factories and in the fields', *Annual Review of Anthropology* 31: 395–417.

Palmer, R. 1977. *Land and Racial Domination in Rhodesia*. Berkeley and Los Angeles, CA: University of California Press

Parry, J. P. 1999. 'Lords of labour: working and shirking in Bhilai', *Contributions to Indian Sociology* 33 (1): 107–40.

2003. 'Nehru's dream and the village "waiting room": long distance labour migrants to a central Indian town', *Contributions to Indian sociology* 37 (1 and 2): 49–81.

and M. Bloch. 1989. 'Introduction', in J. P. Parry and M. Bloch (eds), *Money and the Morality of Exchange*. Cambridge: Cambridge University Press.

Pelkmans, M. 2006. *Defending the Border: identity, religion, and modernity in the Republic of Georgia*. Ithaca: Cornell University Press.

Phimister, 1988. *An Economic and Social History of Zimbabwe 1890–1948: capital accumulation and class struggle*. London: Longman.

Pilossof, R. 2011. *The Unbearable Whiteness of Being: farmers' voices from Zimbabwe*. Harare: Weaver Press.

Polanyi, K. 2001[1944]. *The Great Transformation: the political and economic origins of our time*. Boston, MA: Beacon Press.

Potts, D. 2006. '"Restoring order"? Operation Murambatsvina and the urban crisis in Zimbabwe', *Journal of Southern African Studies* 32 (2): 273–91.
Preston-Whyte, E. 1991. 'Invisible workers: domestic service and the informal economy', in E. Preston-Whyte and C. Rogerson (eds), *South Africa's Informal Economy*. Oxford: Oxford University Press.
Rabe, M. 2006. 'Black workers, fatherhood and South Africa's gold mines', in P. Alexander, M.C. Dawson, and M. Ichharam (eds), *Globalisation and New Identities: a view from the middle*. Johannesburg: Jacana.
Raftopoulos, B. 2003. 'The state in crisis: authoritarian nationalism, selective citizenship and distortions of democracy in Zimbabwe', in A. Hammar, B. Raftopoulos, and S. Jensen (eds), *Zimbabwe's Unfinished Business: rethinking land, state and nation in the context of crisis*. Harare: Weaver Press.
2009. 'The crisis in Zimbabwe, 1998–2008', in B. Raftopoulos and A. S. Mlambo (eds), *Becoming Zimbabwe: a history from the pre-colonial period to 2008*. Harare: Weaver Press.
and I. Phimister. 2004. 'Zimbabwe now: the political economy of crisis and coercion', *Historical Materialism* 12 (4): 355–82.
Rajak, D. 2008. '"Uplift and empower": the market, the gift and Corporate Social Responsibility on South Africa's platinum belt', in G. De Neve, P. Luetchford, J. Pratt, and D. C. Wood (eds), *Hidden Hands in the Market: ethnographies of fair trade, ethical consumption, and corporate social responsibility* (Research in Economic Anthropology, Volume 28). Bingley: Emerald Group Publishing.
Ramphele, M. 1993. *A Bed Called Home: life in the migrant labour hostels of Cape Town*. Edinburgh: Edinburgh University Press.
Ranger, T. 1982. 'The death of Chaminuka: spirit mediums, nationalism and the guerilla war in Zimbabwe', *African Affairs* 81 (324): 349–69.
1985. *Peasant Consciousness and Guerilla War in Zimbabwe: a comparative study*. London: James Currey
1995. *Are We Not Also Men? The Samkange family and African politics in Zimbabwe 1920–64*. London: James Currey.
2004. 'Nationalist historiography, patriotic history and the history of the nation: the struggle over the past in Zimbabwe', *Journal of Southern African Studies* 30 (2): 215–34.
Robins, S. L. 2005. 'Introduction' in S. L. Robins (ed.), *Limits to Liberation after Apartheid: citizenship, governance and culture*. Oxford: James Currey.
Rogerson, C. 1991. 'Home-based enterprises of the urban poor: the case of spazas', in E. Preston-Whyte and C. Rogerson (eds), *South Africa's Informal Economy*. Oxford: Oxford University Press.
Roitman, J. 2004. 'Productivity in the margins: the reconstitution of state power in the Chad basin', in V. Das and D. Poole (eds), *Anthropology in the Margins of the State*. Oxford: James Currey.
Rutherford, B. 2001. *Working on the Margins: black workers, white farmers in post-colonial Zimbabwe*. London: Zed Books.
2003. 'Belonging to the farm(er): farm workers, farmers, and the shifting politics of citizenship', in A. Hammar, B. Raftopoulos, and S. Jensen (eds), *Zimbabwe's Unfinished Business: rethinking land, state and nation in the context of crisis*. Harare: Weaver Press.
2006. 'Borderline projects: Zimbabwean farm workers and (im)mobile resource strategies in South Africa'. Paper presented at the LSE Anthropology Department Africa Seminar, 22 June.

2008. 'An unsettled belonging: Zimbabwean farm workers in Limpopo Province, South Africa', *Journal of Contemporary African Studies* 26 (4): 401–15.

2010. 'Zimbabwean farmworkers in Limpopo Province, South Africa', in J. McGregor and R. Primorac (eds), *Zimbabwe's New Diaspora: displacement and the cultural politics of survival*. Oxford: Berghahn.

2011. 'The uneasy ties of working and belonging: the changing situation for undocumented Zimbabwean migrants in northern South Africa', *Ethnic and Racial Studies* 34 (8): 1303–19.

and L. Addison. 2007. 'Zimbabwean farm workers in northern South Africa', *Review of African Political Economy* 34 (114): 619–35.

Sampson, A. 2005. *Drum: the making of a magazine*. Jeppestown: Jonathan Ball.

Sassen-Koob, S. 1987. 'Issues of core and periphery: labour migration and global restructuring', in J. Henderson and M. Castells (eds), *Global Restructuring and Territorial Development*. London: Sage.

Scarnecchia, T. 1999. 'The mapping of respectability and the transformation of African residential space', in B. Raftopoulos and T. Yoshikuni (eds), *Sites of Struggle: essays in Zimbabwe's urban history*. Harare: Weaver Press.

Scoones, I., N. Marongwe, B. Mavedzenge, J. Mahenehene, F. Murimbarimba, and C. Sukume. 2010. *Zimbabwe's Land Reform: myths and realities*. Woodbridge: James Currey.

Sennett, R. 1998. *The Corrosion of Character: the personal consequences of work in the new capitalism*. London: Norton.

SPT (Solidarity Peace Trust). 2004. *No War in Zimbabwe: an account of the exodus of a nation's people*. 15 November 2004. Port Shepstone: Solidarity Peace Trust.

2005. *Discarding the Filth: Operation Murambatsvina*. Interim report on the Zimbabwean government's 'urban cleansing' and forced eviction campaign May/June 2005. 27 June 2005. Port Shepstone: Solidarity Peace Trust.

Sylvain, R. 2001. 'Bushmen, boers and baasskap: patriarchy and paternalism on Afrikaner farms in the Omaheke region, Namibia', *Journal of Southern African Studies* 27 (4): 717–37.

Tevera, D. and A. Chikanda. 2009. *Migrant Remittances and Household Survival in Zimbabwe*. Southern African Migration Project Migration Policy Series No. 51. Cape Town: Idasa.

Thompson E. P. 1967. 'Time, work-discipline and industrial capitalism', *Past and Present* 38: 56–97.

Tibaijuka, A. K. 2005. *Report of the Fact-Finding Mission to Zimbabwe to Assess the Scope and Impact of Operation Murambatsvina by the UN Special Envoy on Human Settlements Issues in Zimbabwe*. <http://ww2.unhabitat.org/documents/ZimbabweReport.pdf> (Accessed 18 November 2010).

Trapido, S. 1978. 'Landlord and tenant in a colonial economy: the Transvaal 1880–1910', *Journal of Southern African Studies* 5 (1): 26–58.

Turton, D. 2005. 'The meaning of place in a world of movement: lessons from long-term field research in southern Ethiopia', *Journal of Refugee Studies* 18 (3): 258–80.

Ulicki, T. and J. Crush. 2000. 'Gender, farmwork, and women's migration from Lesotho to the new South Africa'. *Canadian Journal of African Studies* 34 (1): 64–79.

Valencius, C. B. 2002. *The Health of the Country: how American settlers understood themselves and their land*. New York, NY: Basic Books.

van Onselen, C. 1976. *Chibaro: African mine labour in Southern Rhodesia 1900–1933*. London: Pluto Press.
1992. 'The social and economic underpinning of paternalism and violence on the maize farms of the South-Western Transvaal', *Journal of Historical Sociology* 5 (2): 127–60.
1996. *The Seed is Mine: the life of Kas Maine, a South African sharecropper 1894–1985*. Johannesburg: Jonathan Ball.
van Schendel, W. 2005. 'Spaces of engagement: how borderlands, illegal flows and territorial states interlock', in I. Abraham and W. van Schendel (eds), *Illicit Flows and Criminal Things: states, borders, and the other side of globalisation*. Bloomington, IN: Indiana University Press.
van Warmelo, N. J. 1942. 'Introduction', in N. J. van Warmelo (ed.), *The Copper Miners of Musina and the Early History of the Zoutpansberg*, Union of South Africa Ethnological Publications Vol. VIII. Pretoria: Union of South Africa Government Printer.
Vaughan, M. 2010. 'The history of romantic love in Sub-Saharan Africa: between interest and emotion', *Proceedings of the British Academy* 167: 1–23.
Veit-Wild, F. 2009. '"Zimbolicious" – the creative potential of linguistic innovation: the case of Shona-English in Zimbabwe', *Journal of Southern African Studies* 35 (3): 683–97.
Ventura, R. 2006. *Underground in Japan*. Manila: Ateneo de Manila University Press.
Vigh, H. 2008. 'Crisis and chronicity: anthropological perspectives on continuous conflict and decline', *Ethnos* 73 (1): 5–24.
von Holdt, K. and E. Webster. 2005. 'Work restructuring and the crisis of social reproduction: a southern perspective', in E. Webster and K. von Holdt (eds), *Beyond the Apartheid Workplace: studies in transition*. Scottsville: University of KwaZulu-Natal Press.
Wagner, R. 1980. 'Zoutpansberg: the dynamics of a hunting frontier, 1848–67', in S. Marks and A. Atmore (eds), *Economy and Society in Pre-Industrial South Africa*. London: Longman.
Waldman, L. 1996. 'Monkey in a spiderweb: the dynamics of farmer control and paternalism', *African Studies* 55 (1): 63–86.
Weber, M. 1992 (1930). *The Protestant Ethic and the Spirit of Capitalism*. London: Routledge.
Werbner, R. 1991. *Tears of the Dead: the social biography of an African family*. Edinburgh: Edinburgh University Press.
West, M. 2002. *The Rise of an African Middle Class: colonial Zimbabwe, 1898–1965*. Bloomington, IN: Indiana University Press.
White, H. 2012. 'A post-Fordist ethnicity: insecurity, authority, and identity in South Africa', *Anthropological Quarterly* 85 (2): 397–428.
Wilson, F. 1972. *Labour in the South African Gold Mines, 1911–1969*. Cambridge: Cambridge University Press.
Wilson, T. D. 2005. *Subsidising Capitalism: brickmakers on the U.S.-Mexican border*. Albany, NY: State University of New York Press.
Wolpe, H. 1972. 'Capitalism and cheap labour-power in South Africa: from segregation to apartheid', *Economy and Society* 1 (4): 425–56.
Worby, E. 1998. 'Tyranny, parody and ethnic polarity: ritual engagements with the state in northwestern Zimbabwe', *Journal of Southern African Studies* 24 (3): 561–78.

2010. 'Address unknown: the temporality of displacement and the ethics of disconnection among Zimbabwean migrants in Johannesburg', *Journal of Southern African Studies* 36 (2): 417–31.

Zinyama, L. M. 2000. 'Who, what, when and why: cross-border movement from Zimbabwe to South Africa', in D. McDonald (ed.), *On Borders: perspectives on international migration in Southern Africa*. New York, NY: St Martin's Press.

Index

abjection, 7, 20, 209
absentee landownership, 76
accommodation, 169, 171, 172, 179; adaptation of, 14, 116, 119–20, 124, 160; farmer, 55; decoration of, 125; minimum standards of, 176; of seasonal workers, 116
Adult Literacy Centre, 145, 173
affirmative parochialism, 53
African National Congress (ANC), 21
African Renaissance (South Africa), 22
African Women's Prayer Union, 142
Afrikaans, language, 164
Afrikaanse Protestante Kerk (Afrikaans Protestant Church), 52, 57
Afrikaners, 30, 46, 54, 72, 174; mythology of, 40–1
AIDS, 149
Alex, a picker, 151–2
Aliens Control Act (South Africa), 100
amnesty for migrants (1988), 13
André, a white farm manager, 60, 164, 165, 166, 175
apartheid, 12, 99, 189; end of, 16; labour migration central to, 7; post-apartheid disillusionment, 22; transition from, 22
aspirations, 31, 142; middle-class, 29, 35–6, 141, 143, 152, 158, 178, 179
autochthony, 21

baboons, guarding crops against, 28, 193

bakkies, 1, 164, 165, 166, 169, 174, 178, 186, 193, 195, 200
bandits *see* labour bandits
bankruptcy, 56, 59
Basotho people, 43
'becoming a man', 65
beer: buying of, 199; confiscation of, 190; drinking of, 154, 155; selling of, 162, 171, 186, 192, 193 (illegal, 118) *see also* drinking groups
Beitbridge border post, 154, 167, 168, 186
Benjamin, a storeman, 30, 33, 143–4, 145, 157, 167, 178
BI-17 permits, 100
Black Economic Empowerment, 172–3
black foremen *see* foremen
black frontmen, promotion of, 49
black managers, 36, 177; of packsheds, 112
black reserves, creation of, 76
black workers, 5, 9–10, 41, 78, 86–7, 99, 110, 161–5, 174, 176, 183; rivalry between, 161
black workforce, employment and domesticity in, 105–36
Boers, 53, 75–6
border economy, of farming, 182–206
border farmers: aggression as driving force of, 53; as counter-insurgents, 14; stories of, 41; role of, in controlling security of territory, 102
border farms and farming, 15, 20, 21, 25, 47; as buffer zones, 71–4, 76, 99; in international economy, 23; rise of, 4; viewed as serious

235

Index

border farms and farming (*cont.*) business, 52–4; wider contexts of, 33 *see also* failure, in border farming *and* success, in border farming
border wars, 14, 69, 72
borders: as place of crisis, 5; cultures of, 67; economies of, 70; governance of, 88, 98; labour economy of, 75–9; policing of, 68–9; porosity of, 107; shaping of, 36; 'special employment' in, 99–103 *see also* policing, of borders
boreholes, 61
bossboy *see* foreman
braai (barbecue), 57, 60, 177, 192
bridewealth, 144
British military operations, 77
buffer zones, 70 *see also* border farms and farming, as buffer zones

Calvinist theology, 52
camaraderie among male pickers, 151
capital flight, 64
casualisation of farming, 6, 107–8
Central Methodist Church (Johannesburg), 141; as place of refuge, 6, 33
chain of command *see* hierarchies
Chewa people, 109
chieftaincy, resurgence of, 21
child labour, 75–6
childcare, 153; during harvest time, 131; women engaged in, 28
children, born to migrant women, 29
China, exports to, 23
Chipo, a picker and businessman, 154, 156, 191, 201–2
ChiShona, language, 30, 32–3, 109, 122, 125, 130, 136, 147, 154
Christianity, 142, 149, 155
Churches in the labour compound, 142, 149, 155; apostolic bishop, 107, 149; Armed for Harvest services, 155; methodology of studying, 31
church services (farmers'), 57

churches, Afrikaans Protestant 52, 57; apostolic, 31; Dutch Reform, 52, 57
cigarettes: selling of, 154, 191; smuggling of, 184, 190
citizenship, Zimbabwean, redefinition of, 18
citrus fruit: production of, 24, 47–8, 60, 62, 90, 145–51; exports of, 22–3; packing of, 112
civilisation, bringing of, 47
class: among migrant workers, 140–5; among pickers, 151–8; class-based status, 12, 138, 142, 143, 145, 163, 178–9; divisions of, 163, 179; models of, in Zimbabwe, 141; reimagined and contested, 36 *see also* middle class
class consciousness in conditions of displacement, 138, 145
coal prospecting on farms, 4, 36, 64, 65, 207–11
commando groups, 72, 73
compound shop, 185, 197
compounds, 9, 27–8, 35; as zone of autonomy, 190; cleaning of, 133; gendered nature of, 128; living in, 30; models of hierarchy in, 168–74; in mining industry, 115–16, 120, 124, 189; residence in, 32; rules of, 29; seasonal arrangements of, 116–22; selfbuilding of, 169; social life of, 31; transience and rootedness in, 115–27
Congo, Democratic Republic of, 6, 183
copper mining, 75, 79–89, 162 *see also* Messina Copper Mine
'core' (hub, hinterland), in Limpopo Valley, 67–71
Cornelius, a driver, 194–6, 197
Corporate Social Responsibility, 6
corporatisation of farms, 49, 160–1, 163, 180
cotton: exports of, 22–3; production of, 38, 45, 47, 60, 62 *see also* ginning of cotton

Index

cows, investment in, 203
credit, a risky practice, 196 *see also* traders, sell on credit
credit groups, 199
currency blackmarket, 204

dams, building of, 42, 48, 64, 65
dancing, 155
Dangarembga, Tsitsi, *Nervous Conditions*, 142
Daniel, a tractor driver, 192
De Beers diamond mine, 104
debt, 7, 20
deportation of Zimbabweans from South Africa, 5, 34, 94, 103, 105, 122, 134, 167
deportation raids, 3, 28, 68, 69, 101, 106, 121
differences among migrants, 12
Dirk, a former farmer, 59–60
disembeddedness, 213–14; of economic relations, 212–13; of land, 216
dismissal of farm workers, 105
displacement of peoples, 5, 7, 140–5; as paradigm, 8; characteristics of, 20–1; in Zimbabwe, 18
diversification of farmers' activities, 63–4
Doctors Without Borders, 5
domestic work *see also* women, as domestic workers
domesticity, 142–3, 156; in compounds, 127–33
drinking groups, 28
drivers, 167
drunkenness, 118
Dutch Reformed Church *see* Nederduits Gereformeerde Kerk

education, 30, 141, 142, 143, 144, 151, 153, 154, 156, 172, 173, 178, 179; in farm context, 129 *see also* university education
electrical goods, use of, 124
electricity, provision of, 51
elephants, hunting of, 43, 75

elites, study of, 143
embedding, of workplace arrangements, 214–15
employment: as incorporation, 26–9; categories of, differences of, 136
English, language, 33, 175
Epstein, A. L. 162–3
Esther, a woman shopkeeper, 185
ethical produce, 160
ethnicity, 152, 153; among pickers, 151–8; as category of difference, 12, 139
EUREPGAP agency, 49
Europeans, seen as different from Africans, 46
Ezekiel, a supervisor, 148, 149

failure in border farming, 34, 38–66 *passim*
Fanakalo, language, 165, 175
Farm Labour Scheme, 90, 91, 92, 99
Farm Watch, 52, 53
farm work: demonisation of, 177; perceived as low status, 137, 178
farm workers, differentiation of, 180
farmers: as jacks-of-all-trades, 50; dress of, 159; identity of, 174
farming equipment, exported from Zimbabwe, 38–9
farming, ideals of, 9
farms: as social centres, 67; as totalitarian institutions, 41; residence on, regulation of, 196
fast track resettlement, in Zimbabwe, 17
favours, dispensing of, 170, 171, 180
fencing *see* security fencing
Flexibility, in white farming, 5, 6, 14, 22–5, 29, 34, 62–5, 114, 136, 166–7, 210, 213, 214, 215, 216; in strategies of workers and compound residents, 143, 186, 197, 198, 199, 200, 205
football, 31, 149, 177, 179; organisation of farm teams, 170–1; team fielded by soldiers, 107

Fordism, 23; racial, 22
foreigners, attacks on, 6
foremen, 36, 60, 181; role of, 100, 161–2, 165
formal economic activity, definition of, 185–91
fragmentation, 207–17 *passim*; and capitalist production, 211–13; of families, 11; of Grootplaas population, 108; of labour, 1–37, 160, 182
furniture: of compound residents, 124, 171; of farmers' homes, 55, 58

Gate 17, 101
gatekeepers, key workers as, 160, 172
gendered nature of compounds, 126–33
George, a picker, 152–3, 209
ginning of cotton, 62
glasses, wearing of, 156
gold mining, 10, 11, 26, 79, 88, 95; remittances of workers in, 203
Granny, a female informant, 32
grapefruit, 112, 140, 145
Green Bombers, 19
Gregory, a pump worker, 128, 191
Grenfell, John Pascoe, 79–80
Grootplaas farm, 2, 21, 160; as lifelong place of employment, 122; as permanent residence, 129; becoming a resident of, 29–34; citrus production of, 23, 47–8; compound shop *see* compound shop; diversity of population of, 109; domestic arrangements in, 106; employment structures of, 140; future prospects for, 207–17; gendered nature of compound, 128; integration of labour into, 27; masculinity in, 139; mix of formal and informal livelihoods in, 183; organisation of work in, 213–14; personal interactions in, 180–1; residential arrangements in, 169;
seen as place of failure, 141; trader activities in, 189, 191–8
guerrilla movements, 13
Guidance, brother of Josiah, 193
guns, 73, 75; 43, 159

Hardship, keeper of employment register, 2–3, 148, 170–1
Hart, Keith, 187
harvest time, 29, 134, 137, 140, 146, 165, 167–8, 169, 186; entertainments during, 107; life in compounds during, 118–19; recruitment for, 2–3
healers, 169 *see also* Sangoma
Henderson Consolidated Corporation, 84
Hendrik, a farmer, 73
hierarchies, 29, 65; models of, in compound, 168–74; nodal points of, 49; of labour, 4, 8, 25, 27, 35, 114–15, 116, 159–81, 175, 214: (gendered, 28); of packshed workers, 113; of traders, 26
hiring and firing, 168
history: genre of, 40; of Afrikaners, 40–1; of settlers, preservation of, 44
HIV/AIDS peer educators, 30, 31–2, 177
Holly, partner of Michael, 128, 129
home-based enterprises, 189–90
homosexual, church ministers, 57
white claims to local history: hospital registers, removal of, 44
Hostel building (Grootplaas), 116–18
Houmoed, name of farm, 39, 43
Housing Committee (Grootplaas), 116
housing, 140; adaptation of *see* accommodation, adaptation of; of farm owners, 55; of compound, 115; illegal, demolished in Zimbabwe, 18 *see also* accommodation
hunting, 43, 177; of elephants, 43; for ivory trade, 75 *see also* poaching of game

Index

hyperinflation in Zimbabwe, 5, 16, 18, 19, 20, 27, 198, 199, 203, 204; responses to, 205
identity documents, 28, 68, 69, 120, 135, 167, 191; produced by farms, 103, 121 see also BI-17 permits
'illegals', 2, 3, 68, 69, 103, 118
immigration, clandestine, control of, 89–99
Immigration Act (South Africa), 93
Incorporation of migrants into farm workforces, 4, 5, 12, 14, 21, 26–9, 214
induna, use of term, 174
informal economy, 6, 23, 205–6; definition of, 182, 185–91
informal livelihood strategies, 182–5
informal trade *see* trade
informalisation, 9, 20–1, 187
informality: concept of, 189; negotiation of, 197; rejection of, 187; usefulness of concept of, 188
inheritance tax, avoidance of, 62
insecurity of labour, 5–9
intercalary figures, 162–3
interviews with informants, 31; organisation of, 33–4
irrigation, 60, 61, 65, 72, 91, 108, 114 *see also* dams
ivory: poaching of, 79; trade in, 75

Jacques, son of Willem, 57–9, 63, 177
Jameson, a picker, 154–8
Jan, a farmer, 38, 46, 53, 56, 74, 100, 102
Jenny, wife of Jameson, 154–6
Jim, a farmer, 51–3, 54
Johannesburg, 6, 8, 33, 78, 123; seen as place of opportunity, 140–1
John, a farmer, 46
Josiah, a shebeen owner, 193
Joyce, a permanent worker, 129–30, 132
justice: on border farms, 35; justice, systems of, 183; permanent workers' versions of, 135

Karanga people, 109
KhaDeki farm, 45, 59
KhaGideon farm, 45
KhaKobus farm, 45
KhaRudi farm, 45, 123
khoro community/compound court, 149, 169–70
kinship networks, 100, 106, 212
Klein Thinus, 57
Kleinplaas estate, 63, 123
Koos founder of Grootplaas, 2, 38, 39, 43, 45–7, 54, 56, 174–7, 181; biography, 43–5
kraals, 115
kukiya-kiya (making do), 7

labour: control of, 71, 85–9, 89–99, 105; economy of, 75–9; free/unfree dichotomy, 79; shortages of, 87, 90, 94; state policy for, 94–9
labour agents, control of, 96–7
labour bandits, 77–9, 88, 97
labour depots of government, 89–94, 96–8; a failure, 93, 98
labour reserves *see* reserves of labour
land claims, 61–5, 210
Land Claims Court, 61
land redistribution, in Zimbabwe, 17
land reform: in South Africa, 24, 61; in Zimbabwe, 19, 42
landmine attacks, 73
landscape: as canvas for development, 61; conceptions of, 47
languages: Afrikaans, 164; ChiShona, 30, 32–3, 109, 122, 125, 130, 136, 147, 154; English, 33, 175; Fanakalo, 165, 175; linguistic complexity of fieldwork, 32–3; multilingual nature of border farms, 110; SeSotho, 33, 177; SiNdebele, 33; soldiers' lack of, 134; TshiVenda, 15, 32–3, 60, 109, 110, 125, 130, 147, 165, 179; XiShangani, 33
latrines in compounds, 116; in mining industry, 87–8
Lemba people, 80
liberalisation of market, 4, 16, 61–5

lice, 120
Limpopo River, 67; crossing of, 1, 76: (risks of, 3); labour context of, 1–37
Limpopo Valley: as buffer zone, 70; as labour recruitment area, 86, 99; as place in own right, 103; characteristics of, 69–70; framing of, 92; rise of agriculture in, 89–94
Lindiwe, a woman shopkeeper, 185
Lindsey, daughter of Michael and Purity, 128
literacy, 167; adult, 173 see also Adult Literacy Centre
living from one day to the next, 27
Louis Trichardt depot, 92, 95, 96, 98
Lovely, niece of Michael, 129

ma-cutters, 140
machismo, 148
magumaguma gangs, 3, 137, 184
MaiJimmy, 192–3, 196, 197, 208
maize, production of, 94
maize meal, 125; distribution of, 186
making do, 40
making ends meet, 36
Makoni, Simba, 207
makwinya (dough balls), 193
malaria, 81, 82–3
Malawi, migrants from, 11, 19
management, in relation to paternalism, 159–81
managerialism: farmers' perspectives on, 174–7; registers of status in, 177–80
managers, white, 55
Manika people, 43
ma-nose brigades, 143
mapermanent, 8, 28, 34, 107, 136, 149, 182, 193, 196, 207, 209, 214; adaptation and rootedness among, 122–7; as aristocracy of labour, 135; investment in compound, 210–11; permits of, 135; remittances of, 198; working context of, 108–15

mapoto, informal relationships, 130, 132
Mapungubwe settlement, 75
Margaret ('Granny'), 130–1; as a Grootplaas resident, 144–5
Marie, wife of Willem, 32, 49, 167
marijuana, 148, 192
marketing boards, 23
marriage, 58–9, 142, 143, 156; farm marriages, 28, 127, 129; ratification of, 129; white weddings, 141
Marula, a black foreman, 2, 27, 109, 122–3, 131, 149, 161–81, 210; activities of, 164–74
Marxism, 11, 12
masalads, 143
masculinities: differences among, 157–8; performed, 148
masculinity: in picking process, 145–58; models of, 139; scholarship on, 139
mashonja (mopane worms), 200
Matabeleland, 15, 33
mbanje, 192 see also marijuana
mealies, 210 see also maize
mediated paternalism, 213–24
Memorandum of Understanding (2004), 102
men, reimagining of, 137–58
Messina see Musina town
Messina Copper Mine, 34, 44, 70, 72, 78, 90, 91, 95, 103; accident books of, 95; labour issues at, 79–89: (recruitment difficulties, 87)
Messina Company, Health Committee of, 82–3, 85, 88
Messina (Transvaal) Development Company, 83, 85
Michael, a personnel manager, 30–1, 32, 33, 125, 128–9, 156, 161–81, 216–17
middle class, 141, 157, 158; as farm workers, 137–58
middle-class aspirations see aspirations, middle-class
migrants: categories of, 21; deaths of, 1 see also undocumented migrants

migration: dependence on, 90; forced, 7; motives for, 17; of labour, 7, 10: (as paradigm, 8; bilateral treaties, 100; scholarship regarding, 12, 26); oscillating, 26; policing of, 74; state policies for, 78, 88–9; viewed in regional context, 77–8 *see also* undocumented migrants
migration routes, 68
militarisation of border areas, 68, 74
military service, of farm workers, 178
military *see* soldiers
mining industry: masculinity in, 139, 150; poor infrastructure provision of, 87; sexual activities in, 148 *see also* copper mining *and* gold mining
misogyny, 147–8
mobile phones, 167, 186
mobility: cross-border, production of, 15–16; differing forms of, 182; in border areas, 99, 109; meaning of, 5–9; of labour, 110, 114; of migrants, restriction of, 133; of status, downward, 137; of white farmers, 40; of women, 128
models of racial difference, ice man / tropical man dichotomy, 46
monogamy, Christian, 142
moonlighting, 190–1
mopaneveld, 1, 48, 77
Movement for Democratic Change (MDC) (Zimbabwe), marginalisation of activists, 17, 19, 152, 157, 207–9; citizenship disqualification of activists, 18
Mozambique, 46; leasing of land in, 63; migrants from, 11, 19; option of relocating farms to, 215; purchase of land in, 4, 64
Mphephu, Chief, 76, 77
Mpho, son of Marula, 123
Mugabe, Robert, 9, 16, 20, 34, 107, 208
muriwo / muroho plant, 125
musanda (headman), 149
musha (rural home), 122, 136, 203

music, anthropological studies in the region, 148, 151; live music gigs on the farms, 107; music on speakers at shebeens, 118, 193; playing of, 171
Musina town, 63, 75, 123, 186; shopping in, 201; wholesale buying in, 197
Mutambara, Arthur, 157

National Constitutional Assembly (Zimbabwe), 17, 154, 157
National Defence Force (South Africa), 68
Ndebele people, 109, 110, 139, 153, 179
Nederduits Gereformeerde Kerk (Dutch Reformed Church), 52, 57
neo-liberalism, 9, 108, 159, 187; in South Africa, 21–6
networks, development of, 15
New Houses (Grootplaas), 116–18, 151, 152, 155
Ngoni state, 26
Njelele River, 89, 90; agriculture in region of, 91, 95
non-contractual relationships, 214, 215
Norman, a senior driver, 129–30
Northern Sotho people, 109

occult economies, 9
Occupatie Wet, 76
Oceana Consolidated Company, 83
Oom *see* Koos
Operation *Murambatsvina* (Zimbabwe), 18, 152, 157
oranges, 112, 122, 140, 157
ownership (of farm), viewed as handicap, 64

packsheds, 63, 115, 122, 155, 192; backlogs in, 147; expenses of, 24; women working in, 110; work in, 111–13
passage, in Limpopo Valley, 67–71

paternalism, 9, 32, 49–50, 136, 159–81, 173, 212; definition of, 159–60; farmers' perspectives on, 174–7; mediated, 213–14; racialised, 41–2; registers of status in, 177–80; reinvention of, 161
patriarchy, 179
Paul, son of Koos, 52, 55, 64
payday market in compound, 200–1
periphery, in Limpopo Valley, 67–71
permanence: production of, 105–36; provisional, 35
permanent workers *see mapermanent*
permits, corporate, issuing of, 102 *see also* identity documents *and* work permits
personnel managers, 30, 36, 112–13, 125, 128, 144, 161–2, 175
Pfumbi people, 109
pickers, 166, 192; male seasonal, 35–6, 138–9; work of, 111 *see also* seasonal workers
picking process, 139, 145–51, 165
picking work, food breaks in, 147
pioneer stories, 39–45, 59, 62, 65, 159, 181
pioneer texts, 42–5
pioneering projects, 47–52
pioneers, 81
'pirates' *see* labour bandits
place making, 127
plantation economies, 100
poaching of game, 174, 190
Polanyi, Karl, *The Great Transformation*, 211–13
police, 191; black, 103; interventions at farms, 176
policing of borders, 35, 74, 82, 88, 103, 167
polocrosse, 56–7, 60
precarity, 29; among workers and their dependents, 4, 9, 25, 26, 29, 37, 66, 124, 126, 131, 132, 196, 197, 216–17; of farmers, 61, 210
precious stones, smuggling of, 190
pregnancy, danger of, 132
prison system, as source of labour, 93, 94, 99

privacy, among farm personnel, 171–2
privatisation, 22
proletarianisation, 12; delayed, 212
promiscuity, 150
prospecting for coal *see* coal prospecting on farms
prostitution, 132, 153
Protestant ethic, 52
Pula, sister of Michael, 129
Purity, partner of Michael, 128

racial division, 165, 176; in workplaces, 32
racism, 42, 177
radio, jamming of stations, 106
railways, building of, 81, 85
rape, 137, 148
recruitment, of labour, 86, 88, 97–8, 100–2, 106, 168; of South African labour, 102; prohibited areas of, 88, 95, 96; state policy for, 98; recruitment on border farms, 100–2, 106
refrigerators in compound houses, 124
regional perpective on crisis and employment, 9–14
relationships, rooting of, 127–35
remittances, 7, 17, 27, 185, 186, 191, 198; as business, 198–205; as goods for resale, 185; non-perishable goods as, 199
repatriation, 93, 96; 'at one's own expense', 94
researchers, farmers suspicious of, 30
reserves of labour, 10, 11, 26, 141, 193, 209, 211, 213, 216
respectability, 156; in Grootplaas context, 143; Zimbabwean ideal of, 139
Restitution Commission (South Africa), 61
retailers in Zimbabwe, stop stocking controlled goods, 19
Rhodesia, migration from, 11, 13, 86, 89, 96
right to reside, 210–11
right-wing politics, 54

Index

rinderpest outbreak, 76
risk, avoidance of, 62, 63, 65
rivalries between kinds of work authority, 29
roads, 1; building of, 81; improvement of, 51, 71
robbery of migrants, 3
roll-call of migrant workers, 2–3
Rudi, son of Koos, 45, 55
rural backgrounds of seasonal workers, 118
rural governance, outsourcing of, 22
Russian Jewish farmers, 95
Rutherford, Blair, 39, 40, 44, 47, 108, 160

Sam, son of Norman and Sarah, 129
Samkange, Grace, 141–2
Samkange, Thomas, 141–2
sangoma (healer), 43
Sarah, wife of Norman, 129, 132
Saul, Michael's brother, a tour guide, 178
savings of workers, 140
screening of rooms, 125
seasonal labour, description of, 111–13
seasonal workers, age range of, 110; differences among, 139–40; remittances of, 198, 199; working context of, 108–15 *see also* pickers
second economy, concept of, 183
security fencing, 1, 69, 73, 74, 137, 208; cutting of, 134; permeability of, 107 *see also* border
self-defence camps of farmers, 52
self-sufficiency, 65; burdens of, 54–61; in border farming, 38–66 *passim*
sending contexts, complexity of, 33
SeSotho, language, 33, 177
Settlement on the border: of workers and dependents, 4, 8, 20, 34, 110, 116, 126, 130, 133, 134, 149, 183, 189, 210, 217; of farmers, 38, 63, 71–2, 74; Messina Mine, 80–1;

control of patterns of residence around, 83, 85; precolonial and early colonial, 61, 75, 77
sex talk among pickers, 147–50
sexuality and sexual relations, 132, 133, 140, 150, 158; construction of stereotype of insatiability, 148; promiscuity, 147–8; taboo on discussion of, 150
Shaka, king, 46
Shangaan people, 109
shebeens, 30, 31, 118, 134, 190, 192, 193, 208
Sheffield, steel workers in, 215
Shona people, 33, 109, 110, 139, 152, 153–4, 179
Simon, a picker, 151–2
SiNdebele, language, 33
Siyanda, wife of Chipo, 154, 156
smallpox, 83
Smith, Ian, 38
smuggling, 174, 183, 187, 188
social authority, non-formal, 183
socialising, time available for, 56–7
sociality of black workers ('socialism'), 46
soldiers in border areas, 1–3, 24, 68, 101, 105, 107, 120, 122, 137, 191; accept sexual favours, 184; as managers of disputes, 135; collusion with, 133–5; in bars, 192; sympathy with, 134; trained in bush tracking, 184
solitude of farmers, 57–8
sons, sent away from farm, 55
Sotho people, 109, 110
South Africa, 4, 143; agriculture in, 24–5; drawing power of towns, 140; government policy in, 6; historical characteristics of, 9–10; northern border area of, 34–5; unemployment in, 23; xenophobia in, 5 *see also* neo-liberalism, in South Africa
Soutspanberg region, 75, 77, 78, 89, 90, 98, 105
sovereignty, of farmers, 102, 105
Soweto, 124

Space, organisation of, on border farms, 8, 21, 27, 115–16, 122, 124, 128, 133, 148, 156, 172, 185–91, 197
spanking as gesture of MDC support, 207–8
spaza shops, 190, 192–3
'special employment', in border zones, 99–103, 105
squatters, black, control of, 83
stereotypes of Africans, 174
stoeps (verandas), building of, 28, 125
stoves, electric, in compound houses, 124
structural adjustment programmes, 6, 203; in Zimbabwe, 15–16, 17
subsistence agriculture, 210
success in border farming, 38–66 *passim*; burdens of, 54–61; ideals of, modified, 61
sugar cane, production of, 63
'Superman', a picker, 148
supermarkets, 176, 178; as customers, 47, 49; corporate social responsibility agendas of, 180; development projects of, 160, 172–3, 176, 177; pressures to change deriving from, 161
supervisors, developing relations with, 149

Takalani, daughter of Marula, 123
Tatelapa pidgin, 165
taxi drivers, 197
taxi services, 193–6, 200
televisions, ownership of, 124
temperance, 144
Temporalities and rhythms, 3, 8, 20, 36, 64, 108, 114, 126, 132, 147, 185, 186, 197, 209, 216, 217
Tendai, a picker, 151
tenure: rights of, 14; security of, 17
Thinus, a farmer, 38–9, 45, 50–1, 54, 56, 73
trade: combined with migration, 182; cross-border, 198

trade: informal, 28 (mapping of, 188); of body cream, 204; of cooking oil, 204; of green soap, 202, 204; of washing powder, 200, 204
traders, 36, 185, 197, 204, 206; cross-border, 107; dependent on taxi drivers, 197; hierarchies of, 26; in border areas, 8; sell on credit, 196
transience, 3–4, 7, 31; in compounds, 115–27
Transvaal Agricultural Union, 102
Transvaal Consolidated Land and Exploration Company, 83
travelling home, 199
trekboers, 40
Tribal Trust Lands, 99
tribalism, 153
Trichardt, Louis, 89, 90
tropical man *see* models of racial difference, ice man / tropical man dichotomy
Tshigidi, a picker, 149
Tshipise region, 105
TshiVenda, language, 15, 32–3, 60, 109, 110, 125, 130, 147, 165, 179
Tsvangirai, Morgan, 19–20, 157, 208
turnover of labour, 31, 115, 196

umbrella companies, use of, 62
Umkhonto we Sizwe, 71, 73, 74
undocumented seasonal workers, 1–2, 3, 69, 101, 105, 120, 121, 197–8; arrests of, 134; special scheme for, 93
unemployment, 27
United African Apostolic Church, 107, 149
United States of America (USA), settler narratives from, 47
university education of farmers, 58, 59
urban backgrounds of seasonal workers, 118

Van Wyk, Gert (former border farmer), 123

vegetable patches in compounds, 125
vehicles of farmers, 55, 58 *see also* bakkies
Venda people, 67, 76, 77, 80, 110, 139, 151, 180; as market traders, 186; cast as inferior, 152; stereotypes of, 153
vigilantism, 52
violence, 5, 20, 49, 153; against farm personnel, 168; against workers, 123; among pickers, 151; attacks on white farmers, 52; in elections, 207; sexual, 131
visas, acquisition of, 106
'visionary' self-understanding of farmers, 46, 50
voortrekkers, 55; history of, 40
vulgar speech, 156–7
vulnerability: of seasonal workers, 106, 120, 121, 208; of women workers, 135
Vusa, a picker, 151–2

wage labour, 70, 191–2, 205, 213; in spaces of crisis, 5
wages, 78, 168; low, 91; minimum, 24, 150, 176, 198; of black workers, 10
'waiters', 146–7, 152
war veterans, 17, 20
water, access to, 50, 51, 64, 118
white farmers, 2, 9, 11, 23, 34, 48, 101, 165, 175; becoming of, 65; fear of police reprisals, 103; interest in mobility, 14; interface with black managers, 36; leave Zimbabwe, 4, 38; perceived as bringers of civilisation, 47; post-apartheid hostility to, 61; self-view of exceptionalism of, 52; social life of, 55, 56; wives of, 32
white farms and farming, 8, 17, 115, 162, 206; narratives of, 39; scholarship regarding, 42
wife of Koos, 43
wife of Paul son of Koos, 185
Willem, (also known as Mpothe), a farmer, son-in-law of Koos, 2, 30, 48–9, 57–9, 65, 72, 73, 123, 165, 167, 168, 175–7, 180, 215, 216
Witwatersrand, 10, 11
Witwatersrand Native Labour Association, 95
wives: job dependence on, 196; of farmers, 32; of permanent workers, 130; seen as lacking ambition, 58; use of rifles by, 73
women: as domestic workers, 109, 129, 132–3; as home makers, 145; as pickers, 131; as reserves of labour, 193; as seasonal workers, 118; as semi-permanent employees, 28–9; as tractor drivers, 51; as traders, 186, 192–3, 200; building domesticity with, 35; in compounds, establishing research relations with, 31–2, 139; in domestic work, 114; in packshed work, 110, 111–12, 132, 140, 150; lack of secure incomes of, 132; livelihood strategies of, 189; maintenance of domestic ideal, 142; relationships of, 107, 118, 127–33, 134, 150; unattached, in compound, 150; vulnerability of, 135, 196
Wonder, an acquaintance of Michael, 128
work: farmers' view of, 54; in picking process, 145–51
work permits, 193
work rhythm, generation of, 147
World Bank, 46

xenophobia, in South Africa, 5–6
XiShangani, language, 33

young people (in white farming families), difficulties of farming life for, 57–8

Zimbabwe: crisis in, 5, 6–7, 8, 9, 14–21, 110, 137–58, 198, 203, 210 (causes of, 16; post-2000, 16–21); elections in, 33, 207–11; historical characteristics of 9–10; migration

246   Index

Zimbabwe (*cont.*)
  from, 4, 15, 16, 17, 33, 53, 54, 163, 177, 204, 212–13; Human Development Index, 20; multiple currency system in, 209; political issues of, 157
Zimbabwean African National Union (Patriotic Front) (ZANU (PF)), 16–17, 18, 38, 208–9; opposition to, 19–20

Zimbabwean migrants, 9, 13, 20, 110, 121, 128, 137, 154, 179, 185; deported from South Africa, 5; disqualified from citizenship, 18
Zimbabwean People's Revolutionary Army (ZIPRA), 13
zones of exception, 71
Zoutspanberg, 75, 83, 88, 94 *see also* Soutspanberg

Titles in the Series

49. MEERA VENKATACHALAM *Slavery, Memory, and Religion in Southeastern Ghana, c.1850–Present*
48. DEREK PETERSON, KODZO GAVUA, and CIRAJ RASSOOL (eds) *The Politics of Heritage in Africa: economies, histories and infrastructures*
47. ILANA VAN WYK *The Universal Church of the Kingdom of God in South Africa: a church of strangers*
46. JOEL CABRITA *Text and Authority in the South African Nazaretha Church*
45. MARLOES JANSON *Islam, Youth, and Modernity in the Gambia: the Tablighi Jama'at*
44. ANDREW BANK and LESLIE J. BANK (eds) *Inside African Anthropology: Monica Wilson and her interpreters*
43. ISAK NIEHAUS *Witchcraft and a Life in the New South Africa*
42. FRASER G. MCNEILL *AIDS, Politics, and Music in South Africa*
41. KRIJN PETERS *War and the Crisis of Youth in Sierra Leone*
40. INSA NOLTE *Obafemi Awolowo and the Making of Remo: the local politics of a Nigerian nationalist*
39. BEN JONES *Beyond the State in Rural Uganda*
38. RAMON SARRÓ *The Politics of Religious Change on the Upper Guinea Coast: iconoclasm done and undone*
37. CHARLES GORE *Art, Performance and Ritual in Benin City*
36. FERDINAND DE JONG *Masquerades of Modernity: power and secrecy in Casamance, Senegal*
35. KAI KRESSE *Philosophising in Mombasa: knowledge, Islam and intellectual practice on the Swahili coast*
34. DAVID PRATTEN *The Man-Leopard Murders: history and society in colonial Nigeria*
33. CAROLA LENTZ *Ethnicity and the Making of History in Northern Ghana*
32. BENJAMIN F. SOARES *Islam and the Prayer Economy: history and authority in a Malian town*
31. COLIN MURRAY and PETER SANDERS *Medicine Murder in Colonial Lesotho: the anatomy of a moral crisis*
30. R. M. DILLEY *Islamic and Caste Knowledge Practices among Haalpulaar'en in Senegal: between mosque and termite mound*
29. BELINDA BOZZOLI *Theatres of Struggle and the End of Apartheid*
28. ELISHA RENNE *Population and Progress in a Yoruba Town*
27. ANTHONY SIMPSON *'Half-London' in Zambia: contested identities in a Catholic mission school*
26. HARRI ENGLUND *From War to Peace on the Mozambique–Malawi Borderland*
25. T. C. MCCASKIE *Asante Identities: history and modernity in an African village 1850–1950*
24. JANET BUJRA *Serving Class: masculinity and the feminisation of domestic service in Tanzania*
23. CHRISTOPHER O. DAVIS *Death in Abeyance: illness and therapy among the Tabwa of Central Africa*
22. DEBORAH JAMES *Songs of the Women Migrants: performance and identity in South Africa*
21. BIRGIT MEYER *Translating the Devil: religion and modernity among the Ewe in Ghana*

20. DAVID MAXWELL *Christians and Chiefs in Zimbabwe: a social history of the Hwesa people c. 1870s–1990s*
19. FIONA D. MACKENZIE *Land, Ecology and Resistance in Kenya, 1880–1952*
18. JANE I. GUYER *An African Niche Economy: farming to feed Ibadan, 1968–88*
17. PHILIP BURNHAM *The Politics of Cultural Difference in Northern Cameroon*
16. GRAHAM FURNISS *Poetry, Prose and Popular Culture in Hausa*
15. C. BAWA YAMBA *Permanent Pilgrims: the role of pilgrimage in the lives of West African Muslims in Sudan*
14. TOM FORREST *The Advance of African Capital: the growth of Nigerian private enterprise*
13. MELISSA LEACH *Rainforest Relations: gender and resource use among the Mende of Gola, Sierra Leone*
12. ISAAC NCUBE MAZONDE *Ranching and Enterprise in Eastern Botswana: a case study of black and white farmers*
11. G. S. EADES *Strangers and Traders: Yoruba migrants, markets and the state in northern Ghana*
10. COLIN MURRAY *Black Mountain: land, class and power in the eastern Orange Free State, 1880s to 1980s*
9. RICHARD WERBNER *Tears of the Dead: the social biography of an African family*
8. RICHARD FARDON *Between God, the Dead and the Wild: Chamba interpretations of religion and ritual*
7. KARIN BARBER *I Could Speak Until Tomorrow:* oriki, *women and the past in a Yoruba town*
6. SUZETTE HEALD *Controlling Anger: the sociology of Gisu violence*
5. GUNTHER SCHLEE *Identities on the Move: clanship and pastoralism in northern Kenya*
4. JOHAN POTTIER *Migrants No More: settlement and survival in Mambwe villages, Zambia*
3. PAUL SPENCER *The Maasai of Matapato: a study of rituals of rebellion*
2. JANE I. GUYER (ed.) *Feeding African Cities: essays in social history*
1. SANDRA T. BARNES *Patrons and Power: creating a political community in metropolitan Lagos*

Printed in Great Britain
by Amazon